U.S. Marine Corps Combat Water Survival

Plus

Marine Physical Fitness Training and Test

Department of Defense

Pentagon
Publishing

MCRP 3-02C

Marine Combat Water Survival

U.S. Marine Corps

PCN 144 000069 00

To Our Readers

Changes: Readers of this publication are encouraged to submit suggestions and changes that will improve it. Recommendations may be sent directly to Commanding General, Marine Corps Combat Development Command, Doctrine Division (C 42), 3300 Russell Road, Suite 318A, Quantico, VA 22134-5021 or by fax to 703-784-2917 (DSN 278-2917) or by E-mail to **morgann@mccdc.usmc.mil**. Recommendations should include the following information:

- Location of change
  - Publication number and title
  - Current page number
  - Paragraph number (if applicable)
  - Line number
  - Figure or table number (if applicable)
- Nature of change
  - Add, delete
  - Proposed new text, preferably double-spaced and typewritten
- Justification and/or source of change

Additional copies: A printed copy of this publication may be obtained from Marine Corps Logistics Base, Albany, GA 31704-5001, by following the instructions in MCBul 5600, *Marine Corps Doctrinal Publications Status.* An electronic copy may be obtained from the Doctrine Division, MCCDC, world wide web home page which is found at the following universal reference locator: **http://www.doctrine.usmc.mil**.

Unless otherwise stated, whenever the masculine gender is used, both men and women are included.

DEPARTMENT OF THE NAVY
Headquarters United States Marine Corps
Washington, D.C. 20380-1775

6 January 2003

FOREWORD

Marine Corps Reference Publication (MCRP) 3-02C, *Marine Combat Water Survival*, provides Marine Corps combat water survival techniques, procedures, and training standards. This publication also teaches Marines to cross water obstacles and perform water rescues correctly and safely.

This publication is the foundation for teaching Marines correct water survival techniques and procedures that are used throughout the Marine combat water survival program (MCWSP). Once an individual or a unit has completed the MCWSP, this publication can be used as a refresher course before water operations.

MCRP 3-02C supersedes Fleet Marine Force Reference Publication (FMFRP) 0-13, *Marine Combat Water Survival*, dated 16 September 1991.

Reviewed and approved this date.

BY DIRECTION OF THE COMMANDANT OF THE MARINE CORPS

EDWARD HANLON, JR.
Lieutenant General, U.S. Marine Corps
Commanding General
Marine Corps Combat Development Command

Publication Control Number: 144 000069 00

MCRP 3-02C Marine Combat Water Survival

Table of Contents

Chapter 1. Survival at Sea

Chapter 2. Water Rescues

Chapter 3. Treatment of Casualties and Avoidance of Dangerous Marine Life

Chapter 4. Negotiating Water Obstacles

Chapter 5. Fording Waterways

Appendix. Knot Tying

Throughout history, water has posed special challenges to Marines and Sailors during times of both peace and war. Therefore, the inherent dangers associated with waterborne operations demand that Marines and Sailors receive proper water survival training. Combat units that are confident in their ability to work in and around water can use the water to their advantage in combat, and history is filled with examples of how the proper preparation or training for survival in water reduced or averted disaster.

USS *Indianapolis*
On Sunday, 29 July 29 1945, the heavy cruiser USS *Indianapolis* (the *Indy*) was en route to the Philippine Sea. Shortly before midnight (about 39 hours out of port), the *Indy*, running blacked out and unescorted, was rocked by two explosions on her starboard side. With communications smashed, the ship could not signal its distress and sank within 15 minutes.

Three life rafts and a floater net supported a few survivors, but the rest of the survivors drifted about, held up by rubber life belts or Mae Wests. About 60 seamen died the first night. Survivors assumed the ship would be reported overdue in Leyte, and that they would be rescued within 2 days. By Monday evening, however, panic began to set in as some life jackets lost their buoyancy from the long immersion. Some men even fought over life jackets, which resulted in at least 25 deaths. No one dared sleep for fear of losing his jacket.

Throughout the next several days, in-transit aircraft flew nearby without spotting the desperate seamen. As best they could, the men kept together, some tying long ropes to each other, floating like corks on a net.

Not until late Thursday morning, 3 1/2 days after the ship sank, were the men discovered. Luckily, a plane on a routine flight over the area spotted the survivors. When ships picked them up that night, the survivors learned they had never been reported overdue. Every one of the 1,196-man crew was a casualty; 880 crewmembers were listed as dead or missing. Although many lives were lost, the innovative and expedient use of flotation devices and float techniques employed by the survivors helped save hundreds of lives.

USS *America*
On Thanksgiving eve, 23 November 1995, the USS *America* made its way through the Arabian Sea. Twenty-year-old, Marine Lance Corporal (LCpl) Zachary Mayo was unable to sleep and, wanting some fresh air, made his way onto an open-air platform near the aircraft hangar bay, which was three levels below the sleeping quarters. While he was on the platform, the ship veered suddenly, throwing LCpl Mayo through the platform's protective bars and into the sea, 30 feet below.

Frantic, LCpl Mayo called out in vain to the watchmen on the flight deck, which was 64 feet above him. It soon became clear to him that the USS *America* would keep its course into the Gulf of Oman until his absence was discovered at morning muster. The LCpl took a moment to consider his situation. Since land was at least 100 miles away, swimming was suicide; he would have to stay afloat until a search party found him.

Using the techniques he had learned during combat water survival training, LCpl Mayo made a flotation device out of his coveralls and tried to relax. Meanwhile, business continued as usual aboard the USS *America*. Since LCpl Mayo was on special assignment with the hazardous materials division, his absence went unnoticed until a petty officer asked several Sailors if they had seen their shipmate recently. By the time a roll call had been completed, LCpl Mayo had been adrift at sea for over 24 hours. Although three, fixed-wing Viking aircraft were deployed to search for LCpl Mayo, most people aboard ship feared the worst.

After 34 hours at sea, LCpl Mayo was discovered by fishermen on a Pakistani fishing boat. LCpl Mayo's survival is a testament not only to his incredible physical courage, but also to the soundness of the lifesaving training and techniques he received during combat water survival training.

Chapter 1
Survival at Sea

As a Marine, you face a variety of potential water emergencies whenever you cross expanses of water: ships, watercraft, and amphibious assault vehicles (AAVs) can sink; aircraft can crash into the sea; or you can accidentally fall into the water. However, there are some basic precautionary measures you can take to protect your safety and reduce your chances of becoming a water casualty. Determine the following information as soon as you board any type of vessel. Your knowing the following information may save your life or the life of your fellow shipmates.

- How many life preservers and lifeboats/rafts are on board?
- Where are the life preservers and lifeboats/rafts located?
- What type of unit survival equipment is on board?
- Are individual survival kits issued to each person on board?
- How much food, water, and medicine do the survival kits contain? When was the last time the contents were inspected for proper quantities and shelf life expiration?
- Is there sufficient survival equipment available for the number of personnel?
- How many other personnel are there on board, and where are they located?
- What are the egress procedures for the ship, boat, watercraft, AAV, or aircraft?

Abandoning Ship

When you embark on a Navy ship, you will receive abandoning ship instructions from Navy personnel. If given the order to abandon ship, report to your designated assembly area and put on a life preserver. DO NOT inflate the life preserver until you are clear of the ship. Torn life preservers will not inflate and inflated life preservers can block you, and those behind you, from exiting the ship. A flotation device that has been inflated may also burst if you jump from a significant height. See pages 1-15 through 1-35 for staying afloat with and without a life preserver.

DO NOT remove your clothing, boots, or shoes before abandoning ship. Your trousers and blouse may be the only flotation devices available if your life preserver is faulty or becomes damaged, and your clothes can provide some insulation from the cold water. However, remove your soft cover and place it in a cargo pocket for later use. The soft cover is both lightweight and a good protection against sunburn caused by the sun's rays reflecting off the water.

Jettisoning Equipment

Equipment should be kept properly packed and waterproofed in case you have to abandon ship. If entering the water from a height greater than 30 feet, wearing your equipment (e.g., pack, helmet, gas mask) could cause injury. Upon impact with the water, the helmet will "cup" air inside of it. The chin strap may also create a "hanging effect" as you submerge from the force of the fall. Therefore, you should remove your helmet and gas mask before abandoning ship.

If you are unable to maintain buoyancy due to the amount of equipment secured to your pack and body, then jettisoning some of your equipment may become necessary. Equipment that you should always retain includes canteens of freshwater, first aid kit, soft cover, and survival kit. The survival kit should include first aid items, water purification tablets or drops, fire starting equipment, signalling items (e.g., flashlight, strobe light, chemlights), food procurement items, and shelter items. Other items in a survival kit include sunburn lotion and lip balm, knife, goggles/sunglasses, plastic bag, matches and lighter, mirror. See Marine Corps Reference Publication (MCRP) 3-02F, *Survival*, for a detailed list of survival items and applications.

Abandoning Ship Technique

When abandoning ship, safety considerations must be observed. Use the following technique when abandoning ship without your combat gear:

Place your hands on their opposite shoulders, forming a criss-cross pattern.

Step to the edge of the ship's deck and check the water below for debris or survivors. If the water is clear, look straight ahead and prepare to jump. If the water is not clear, move to another location.

NOTE: DO NOT hold your nose as you abandon ship. If you do hold your nose, the force of impact into the water could jar your arm and hand and cause you to break your nose.

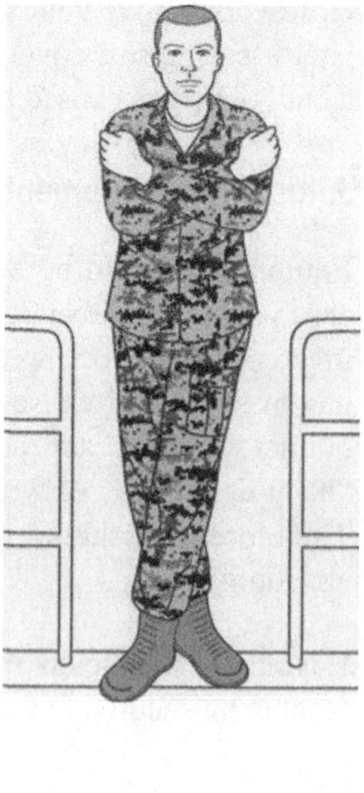

Step off the side of the ship with a smooth, 30-inch stride. DO NOT DIVE OFF THE SHIP. DO NOT LOOK DOWN AT THE WATER. LOOK STRAIGHT AHEAD. Looking down at the water can render you unconscious or cause injuries upon impact.

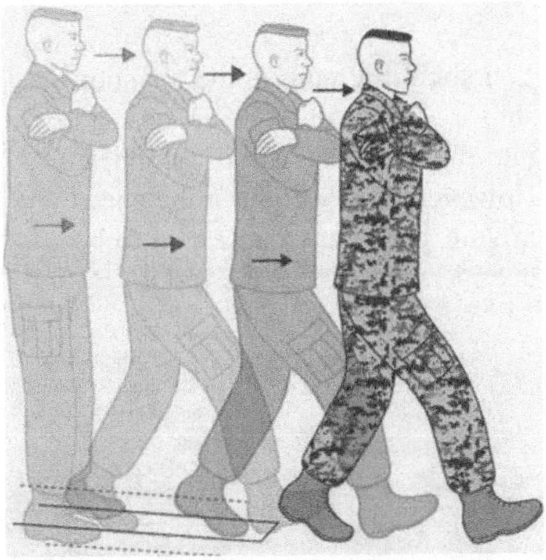

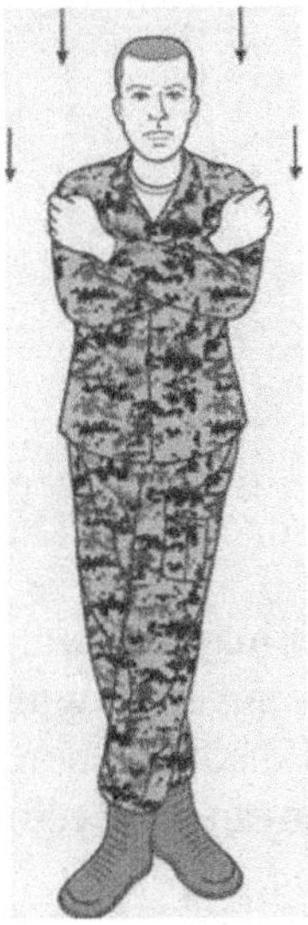

Bring your trailing leg forward during the fall. Cross your trailing leg behind your leading leg.

Keep your head parallel to the water's surface until hitting the water.

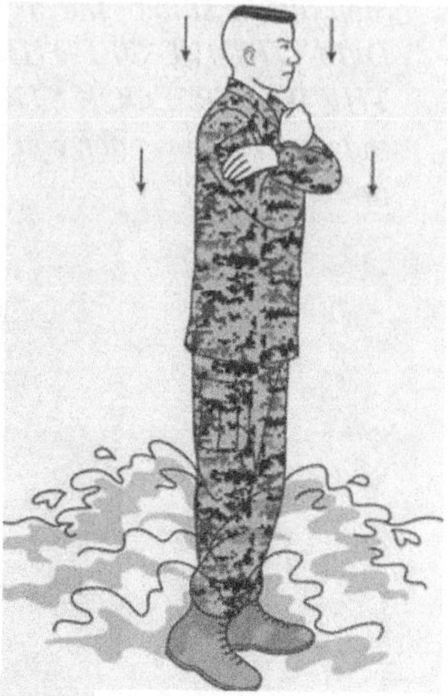

You should remain in the abandon ship position until your descent into the water has almost stopped. However, the weight imbalances in your body may cause you to be in a "J" shape under the water

Once your downward motion has ceased, your feet may be parallel with the ocean bottom or you may be nearly inverted with your feet over your head. To counteract potential disorientation, you should pause briefly and allow the natural buoyancy of your torso to bring your body to a nearly upright position.

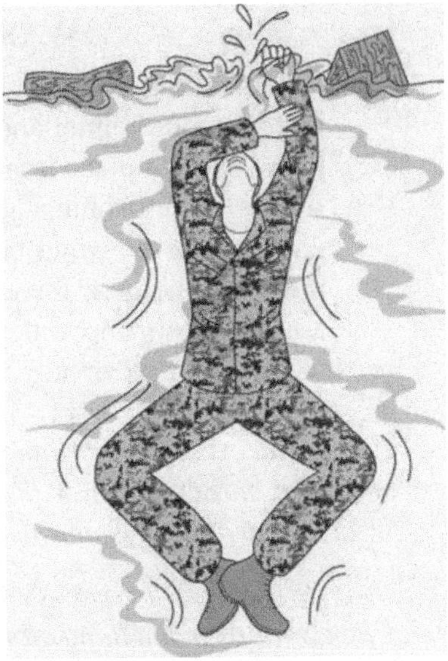

Floating debris can cause hazards. Therefore, you should swim upward, extending one arm (hand is shaped as a fist) upward to feel for obstructions. If you encounter debris, try to push it away or surface in a different location.

Swim away from the ship. DO NOT LOOK BACK AT THE SHIP. Looking back could slow your movement away from the area. Remember, your objective is to leave the area as quickly as possible because—

- Equipment and debris may be falling from or spilling out of the ship.
- Additional casualties can occur if individuals abandoning the ship fall on top of swimmers already in the water.
- Swimmers close to the sinking ship may get pulled underneath the water by the suctioning effect of the ship as it goes under.

Modified Abandoning Ship Technique

When abandoning ship while wearing full combat gear (weapon, helmet, and a properly waterproofed pack), safety considerations must be observed. The modified abandoning ship technique is used while wearing full combat gear and exiting from a height less than 30 feet.

WARNING

When exiting from a height greater than 30 feet, remove your helmet and pack. Fasten the helmet to the pack or place it inside the pack before jettisoning. If jettisoning gear from the ship or aircraft, check the water below for survivors before throwing the gear forward of the intended jump area. Once in the water, retrieve your gear and swim out of the area.

Place your weapon over one shoulder, muzzle down, with the weapon parallel to your side.

Place your arm and hand along the weapon and hold it to your side.

Take your free hand and place it on top of your helmet to prevent neck/spinal injury from the force of the water pulling upward on the helmet as your body enters the water.

NOTE: An alternate method is to place your weapon over one shoulder, muzzle down, with the weapon parallel to your side. Reach across your body and grasp the sling of your weapon and hold it to your body. Take your free hand and place it on top of your helmet.

Step to the edge of the platform and check the water below for debris and survivors. If the water is clear, look straight ahead and prepare to jump. If the water is not clear, move to another location.

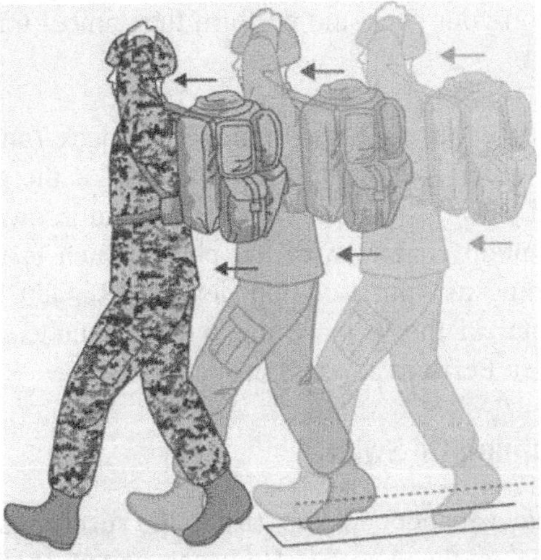

Step off the side of the platform with a smooth, 30-inch stride. DO NOT DIVE OFF THE SHIP, DO NOT LOOK DOWN AT THE WATER, LOOK STRAIGHT AHEAD. Looking down at the water can render you unconscious or cause injury upon impact.

Bring your trailing leg forward during the fall. Cross your trailing leg behind your leading leg.

Keep your head parallel to the water's surface until hitting the water.

You should remain in the modified abandoning ship position until your descent into the water has almost stopped. The buoyancy of a properly waterproofed pack will immediately pull you to the surface. Once you break the water's surface, unsling your weapon and loop the sling over your head. You may have to seesaw the sling so as to ensure the sling passes between your helmet and pack. Once over your head, loop the sling around your neck with the weapon aligned in the center of your body, muzzle down. Lean back on your pack and perform the combat travel stroke to exit the area.

> *NOTE:* You may also remove the pack (unsnap the quick-release strap on one side of the pack, and pass the strap to your other hand in order to maintain contact with the pack, which is functioning as your flotation device). Remain horizontal in the water with your pack under your chest. Perform a breast stroke kick.

Surface Burning Oil Swim

After you have abandoned ship, rise to the surface using the techniques shown on the page 1-6. However, you must remember that fuel from sinking ships or downed aircraft will float on the surface of the water. Therefore, you must move clear of the floating fuel by swimming away from the ship or aircraft as soon as possible. Either swim upwind (into the wind) of the ship/aircraft or swim against the current. Either method allows you to move away from the fuel and the wind/current will push the fuel past you. To properly execute a surface burning oil swim—

Extend your arms overhead as far as possible.

Wave your arms back and forth vigorously to splash a hole while moving upward.

Splash as long as possible to push burning fuel away from the surfacing area.

Use your arms and hands to sweep away fuel and debris.

Kick your legs in a constant breast stroke kick.

Extend your arms (palms outward) forward on the surface, arms shoulder-width apart.

Pull your hands in and back toward the chest.

Stop your hands in front of your face and rotate them so that your palms face forward (roughly halfway out of the water).

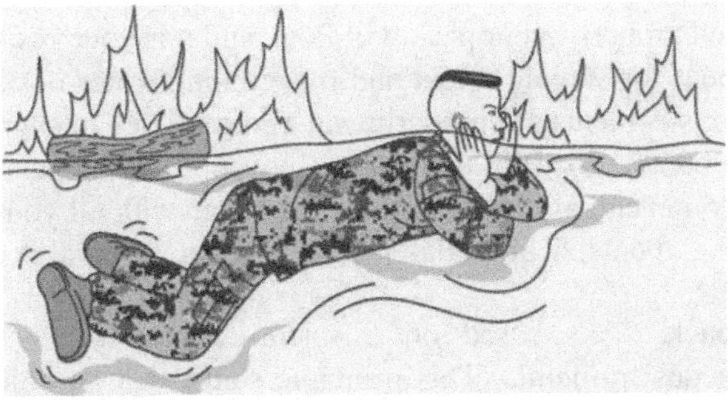

Sweep your arms forward to a full extension at the shoulder width. This splashes debris, oil, or burning liquids aside. To reduce the chance of fatigue, use two short splashes to the front to extend the path.

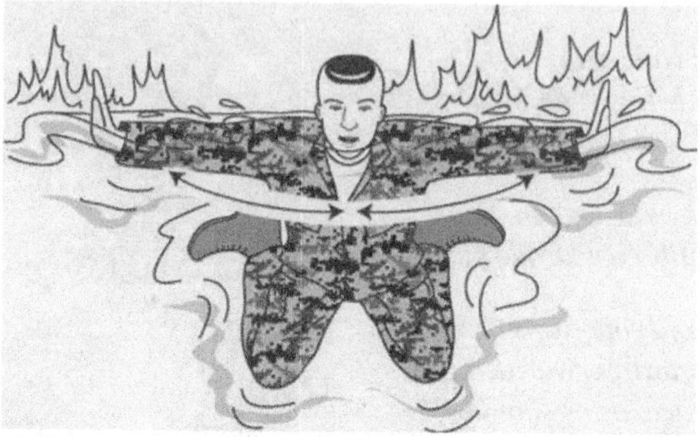

Repeat the preceding step as necessary while swimming clear of the area.

Surviving With a Pack

If packed properly, your pack will float, and it is your key piece of equipment for staying afloat and overcoming water obstacles. If the pack's contents are properly waterproofed, it can support you (with a combat load) in the water. Buoyed up by a waterproofed pack, you eventually emerge from the water with all your equipment (e.g., boots, helmet, flak jacket, weapon, survival items).

Your pack floats based on a scientific principle known as Archimedes' principle. This principle states that an object submerged in a liquid is buoyed up by a force equal to the weight of the liquid displaced (pushed aside) by the object. If the weight of the displaced liquid is greater than the weight of the object, the object floats. If the weight of the displaced liquid is less than the weight of the object, the object sinks. For example, a machine gun sinks in the water, but it still weighs less in the water than it does on land. Even though a machine gun sinks, it is still buoyed up by a force equal to the weight of the water it displaces. For this reason, you should not try to hold yourself or your equipment any higher out of the water than they would naturally float; doing so wastes both energy and body heat.

Preparing Equipment

Before packing your pack, you must prepare your gear/equipment. Tape or pad all sharp edges and equipment corners. Ideally, your gear/equipment is placed in plastic bags and the plastic bag is then placed inside the standard-issued, rubberized, waterproof bag. Waterproof bags are not completely water tight; henceforth, the added protection of first wrapping the gear/equipment in a plastic bag, then placing the plastic bag inside the waterproof bag. Large plastic bags (e.g., trash bags) work well for bulky equipment (e.g., sleeping bags, field jackets, shelter halves, gas masks). Small plastic bags work best for small items (e.g., shaving gear).

> *Note:* If a gas mask must be carried outside the pack, cover it with a waterproof bag.

You will need the following items, which are available through the supply system, to prepare your equipment for packing:

- Waterproof bags:
  Pistol bag, plastic size #1 (8" x 18").
  Rifle bag, plastic size #2 (10" x 56").
  Machine gun bag, plastic size #3 (15" x 56").
  Multipurpose bag, cover, plastic size #4 (20" x 84").

- Panama work vest (life preserver).
- Small plastic bag, self sealing (6" x 6" and 12" x 12").
- Riggers tape, olive drab (2 1/2").
- Willie peter bag, waterproof clothing.

Tying Waterproof/Plastic Bags

Try to remove the excess air from the waterproof/plastic bag before securing the bag's opening. If filled with air, the bag can burst if pressed from the outside. Perform the following steps to tie a waterproof/plastic bag:

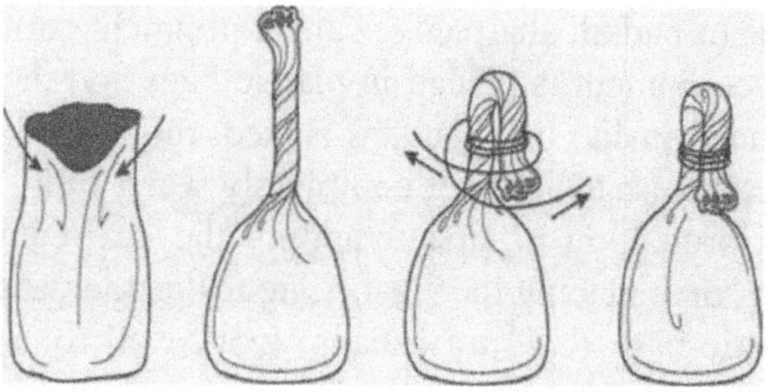

Packing the Pack

Place filled and tied bags inside the pack. Carefully handle sharp items (e.g., tent pegs, poles) to prevent puncturing the bag. Place items in the pack in order of expected use. Close the pack and its compartments. Attach sleeping mats or bags as high as possible on the outside of the pack.

Swimming With the Pack

A useful technique for propelling yourself forward while wearing a pack is the combat travel stroke. This technique allows you to float nearly horizontal and to propel yourself forward with bicycle-style kicks and breast stroke-style arm sweeps. To execute the combat travel stroke—

Body position
Keep the upper part of your body prone to the water, let your legs dangle horizontally, and keep your face up.

Arm action
Extend your hands out in front of your waist. Sweep your arms slightly downward and back 90 degrees to propel your body through the water. Move your hands back to the front of your waist. Repeat.

Leg action
Continuously move your legs in a bicycle-like movement, bringing your knees up high and step out.

Breathing
Keep your face out of the water during the stroke, and breathe freely.

Staying Afloat With a Life Preserver

The best form of flotation is to find any kind of floating object that will keep you and your equipment out of the water or minimize your exposure to the water. Life preservers are the best method, as they allow you to wear your clothes for heat retention and sunburn prevention. Marines use two basic classes of life preservers: inherently buoyant life preservers and inflatable life preservers.

Inherently Buoyant Life Preservers

Inherently buoyant life preservers are either vest-type (worn like a jacket) or yoke-type (worn around the neck). The preserver's outer envelope is either a cotton or water resistant material that encloses a removable fibrous glass or plastic foam filling.

In the Marine Corps, the most common type of inherently buoyant life preserver is the vest-type with collar, known as the kapock preserver. The kapock consists of collar straps, upper front chest straps, leg straps, and waist drawstrings that secure the preserver to you. The leg straps, which are fitted on both sides of the life preserver, ensure that the preserver remains around your chest while you are in the water. A chest strap is attached to the life preserver to facilitate lifting you out of the water. The strap can also be attached to other survivors or to lifeboats to reduce the fatigue that results from holding onto a floating/secured object by hand.

Inflatable Life Preservers

Marine Corps aircraft and AAVs have inflatable life preservers on board. The inflatable life preserver issued to Marines is known as the LPP (life preserver personal). The LPP is capable of both oral inflation and CO_2 cartridge inflation. The LPP consists of buoyancy chambers, CO_2 inflator, and an oral inflation tube. The buoyancy chamber is deck gray in color and is made from a neoprene-coated nylon fabric.

Inflatable life preservers must be stored in a cool, dry place. Heat, moisture, and light cause deterioration of the life preserver material. Do not stow CO_2 cylinders near steam lines or radiators. Heat can increase the pressure inside the cylinders causing them to explode. Avoid sharp edges in stowage. Sharp edges increase wear and tear on the life preservers and may also puncture inflatable buoyancy chambers.

> **CAUTION:** Do not inflate the life preserver until you are clear of the aircraft, ship, or vehicle. Torn life preservers will not inflate and inflated life preservers can block you, and those behind you, from exiting the aircraft, ship, or vehicle.

To don and adjust the LPP—

Remove the LPP from its storage container.

Fasten the belt fasteners in front with the pouch in the rear.

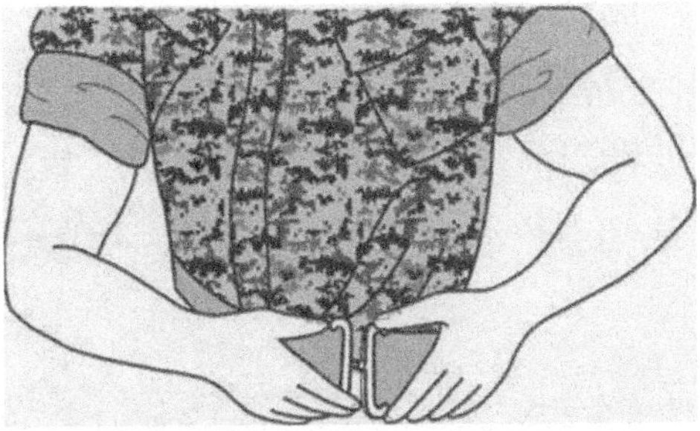

Adjust the belt to fit; secure any excess belt by mating the hook and pile tape.

Rotate the pouch to the front and re-adjust the belt if necessary.

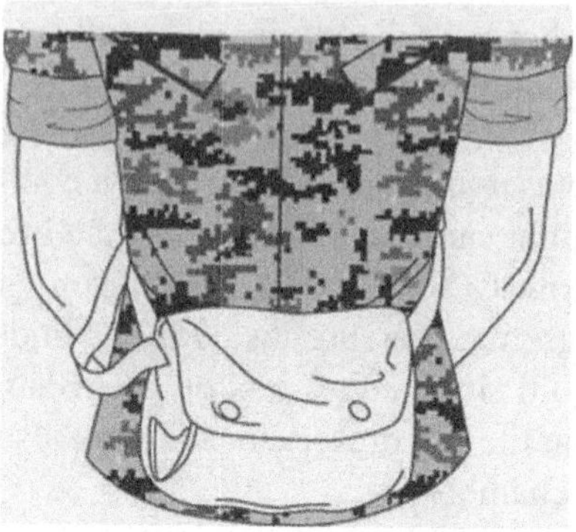

Open the snap fasteners on the pouch and unfold the life preserver.

Place the deflated life preserver over your head.

Place the storage container into the pouch after donning the life preserver.

Lift the lower end of the life preserver out of the pouch.

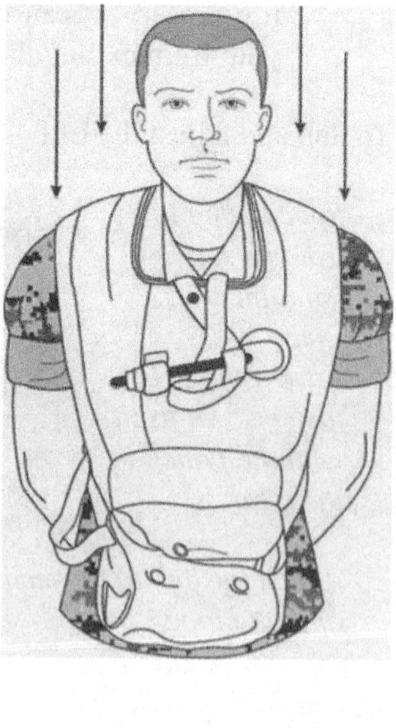

Inflate the life preserver by pulling on the lanyard attached to the CO_2 inflation valve or by blowing on the end of the oral inflation valve.

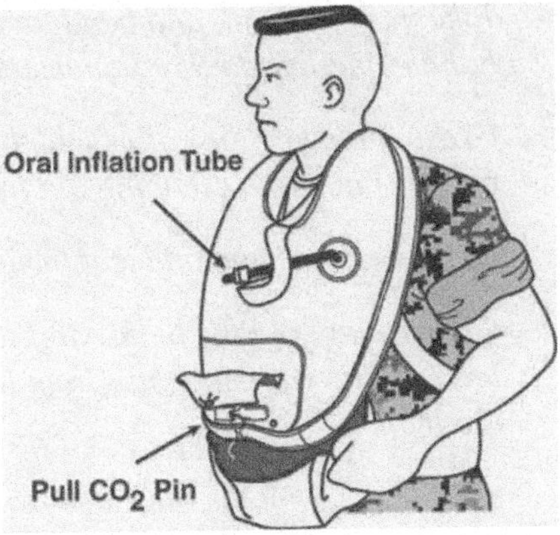

Staying Afloat Without a Life Preserver

You may be in open water without any floating objects or a life preserver to help you survive. If so, uniform blouses and trousers can be made into expedient flotation devices.

Floating With an Inflated Blouse

It is possible to float by a bubble of air trapped in the shoulders of your blouse. The air rises to the back and shoulders of the blouse and supports you at the water's surface. An inflated blouse is also a temporary flotation device used by weaker swimmers while trying to remove their trousers. There is a primary and an alternate way to create a bubble of trapped air in a blouse—

Primary Method
Turn the collar inside the blouse to help create a seal.

Unbutton top button and pull collar around mouth and nose.

Take a deep breath and bend forward slightly at the waist. Exhale one-half to three-quarters of a breath into the blouse.

Grasp and twist the collar with one hand to create a seal, this prevents air from escaping out from the collar.

Use your free hand and feet to stroke and kick to the surface.

Gather and hold the blouse tightly at the collar and stomach level to prevent the blouse from losing air if it floats up too high.

Splash water on the blouse periodically to prevent the material from drying, dry material allows air to escape.

Repeat inflation as required.

Alternate Method
Turn the collar inside the blouse to help create a seal.

Unbutton the second button from the top.

Take a deep breath and bend forward slightly at the waist.

Place your mouth and nose inside the hole created by the open button and exhale one-half to three-quarters of a breath into the blouse.

Grasp material at the unbutton portion and pull downward.

Use your free hand and feet to stroke and kick to the surface.

Splash water on the blouse periodically to prevent the material from drying, dry material allows air to escape.

Repeat inflation as required.

Floating With Inflated Trousers

In warm water, trousers can be used as a primary expedient flotation device. However, in cold water, submerging your head to remove and inflate your trousers results in heat and energy losses that negate the benefit of using the trousers as a flotation device. Once your trousers are inflated, you float motionlessly as if wearing a life preserver. If needed, assume the heat escape lessening posture (known as the HELP position, see page 1-37) to slow heat loss. As trousers dry, air leaks out of the legs. To slow this process, occasionally splash water on the fabric. Reinflate trousers as needed.

Sling Method

The sling method works if you are a strong swimmer or naturally very buoyant. Take the following steps to inflate trousers using the sling method:

Take a deep breath, bend over, and remove your boots.

NOTE: Retain your boots. Tie the boot laces together and suspend the boots from your blouse or hang them around your neck so that they rest on your chest.

Remove your trousers. Button or zip the trouser's fly closed. This allows you to control airflow.

Tie the bottoms of the trouser legs in a square knot. (The appendix illustrates various knots.)

Ensure that the front (fly) of the trousers faces you.

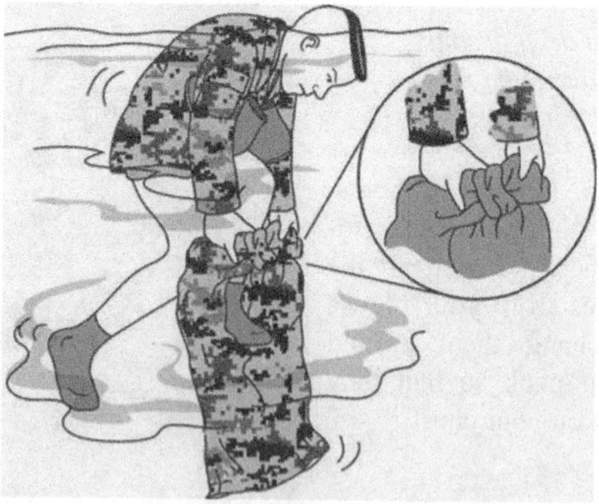

Hold the trousers above the water's surface and behind your head. Grasp both sides of the waistband and open with both hands.

Kick strongly to stay on top of the water while slinging the trousers overhead in order to trap air into them.

Once the waistband is submerged in the water, air is trapped in the legs.

Hold and seal the waistband underwater.

Slip the inflated legs over your head. Hold the waistband in toward your chest, the fly facing your body. To prevent air from escaping from the trousers, seal the waistband by either folding it or twisting it.

Lie back and relax, resting the back of your neck against the knot.

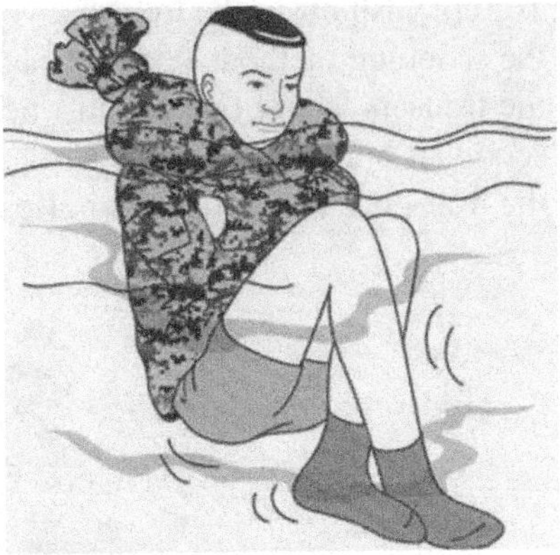

Splash water on the trousers periodically to prevent the material from drying. Dry material allows air to escape.

To replenish air in the trousers, you will use a technique known as the scooping method. With one hand on the open waistband, extend the trousers in front of you just below the surface of the water and scoop air bubbles with your free hand into the open waistband until the trousers have sufficient air. Repeat as necessary.

Splash Method

The splash method is an alternative to the sling method. As with the sling method, you must kick strongly to remain at the surface. To inflate trousers using the splash method, perform the following:

Take a deep breath, bend over, and remove your boots.

NOTE: Retain your boots. Tie the boot laces together and suspend the boots from your blouse or hang them around your neck so that they rest on your chest.

Remove your trousers. Button or zip the trouser fly closed. This allows you to control airflow.

Tie the bottoms of the trouser legs in a square knot. (The appendix illustrates various knots.)

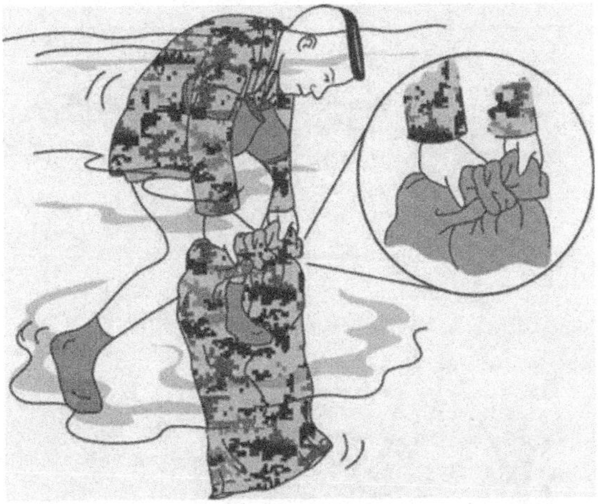

Ensure that the front (fly) of the trousers faces you.

Hold the trousers at the water's surface out in front of you by the waistband with the fly up.

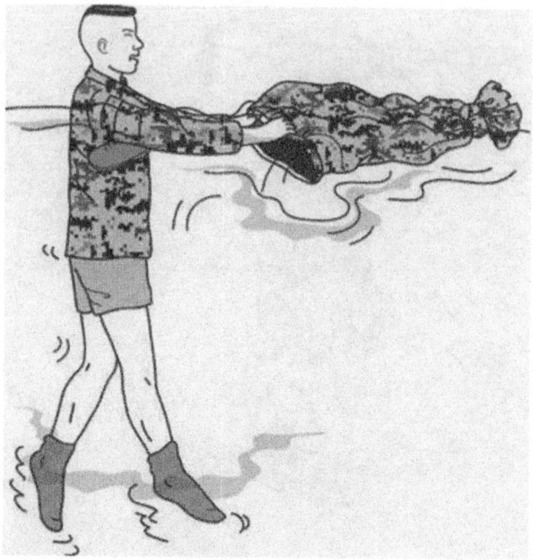

Grasp the waistband at the surface with one hand. Insert your free hand into the waistband, palm down.

Flutter your hand rapidly to create bubbles. This sends a mixture of water and air bubbles into the trousers. The water passes through the fabric. The air remains trapped in the legs.

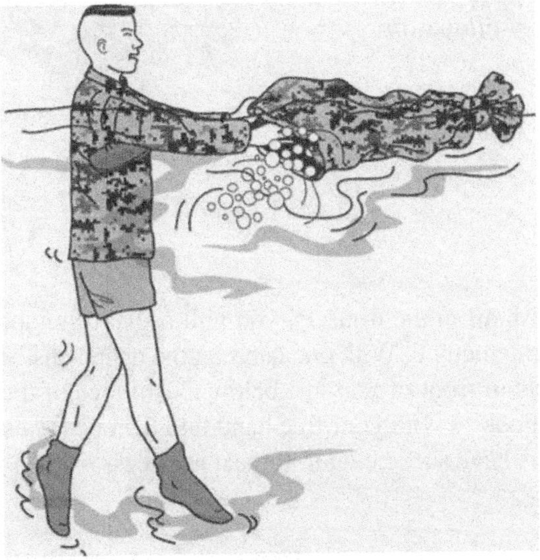

Hold and seal the waistband underwater.

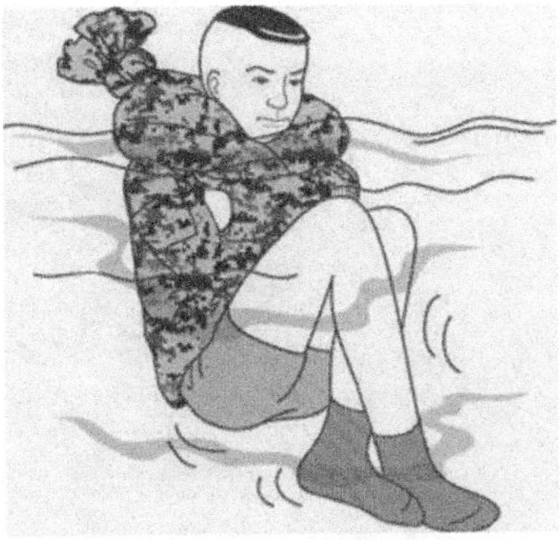

Slip the inflated legs over your head. Hold the waistband in toward your chest, the fly facing your body. To prevent air from escaping from the trousers, seal the waistband by either folding it or twisting it.

Lie back and relax, resting the back of your neck against the knot.

Splash water on the trousers periodically to prevent the material from drying. Dry material allows air to escape.

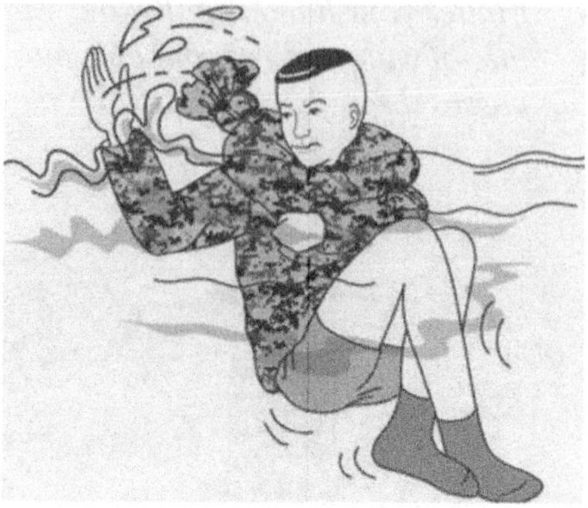

To replenish air in the trousers, you will use a technique known as the scooping method. With one hand on the open waistband, extend the trousers in front of you just below the surface of the water and scoop air bubbles with your free hand into the open waistband until the trousers have sufficient air. Repeat as necessary.

Blow Method

The blow method is an alternative to the sling method. Use the blow method if you are a weak swimmer. Take the following steps to inflate trousers using the blow method:

Take a deep breath, bend over, and remove your boots.

NOTE: Retain your boots. Tie the boot laces together and suspend the boots from your blouse or hang them around your neck so that they rest on your chest.

Remove your trousers. Button or zip the trouser fly closed. This allows you to control airflow.

Tie the bottoms of the trouser legs in a square knot. (The appendix illustrates various knots.)

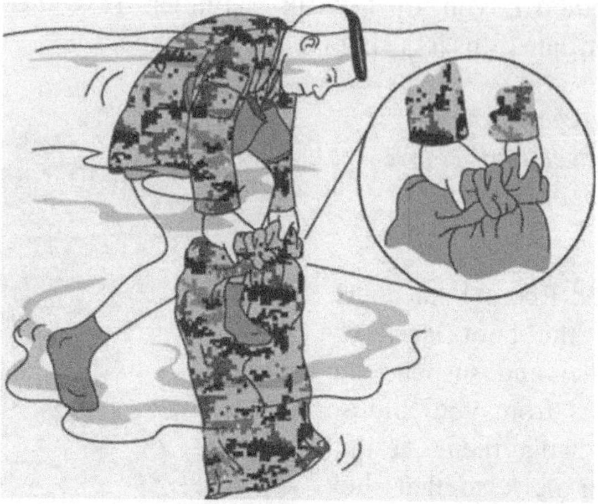

Ensure that the front (fly) of the trousers faces you.

Hold the trousers at the water's surface. Grasp both sides of the waistband and open with both hands.

Take a deep breath.

Drop 2 feet below the water's surface, pulling the waistband underwater.

Hold the waistband open with both hands and blow air into the trousers.

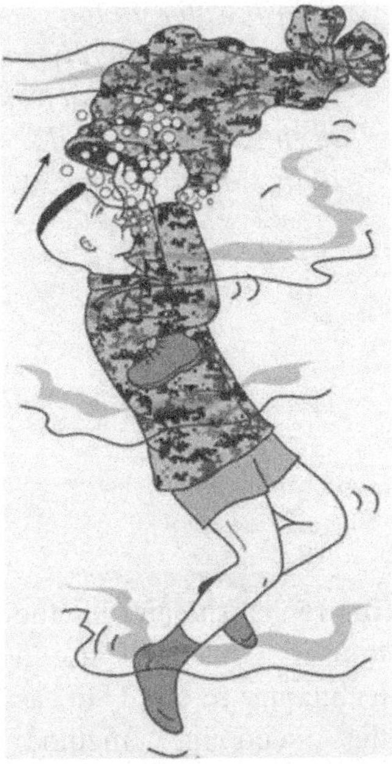

To fill the trousers with air, surface while keeping the waistband underwater, breathe in again, drop below the water's surface, and blow air into the trousers. Repeat these steps until the trousers are filled sufficiently. Once trousers are filled—

Hold the waistband underwater. Twist and pinch it off.

Slip the inflated legs over your head. Hold the waistband in toward your chest, the fly facing your body. To prevent air from escaping from the trousers, seal the waistband by either folding it or twisting it.

Lie back and relax, resting the back of your neck against the knot.

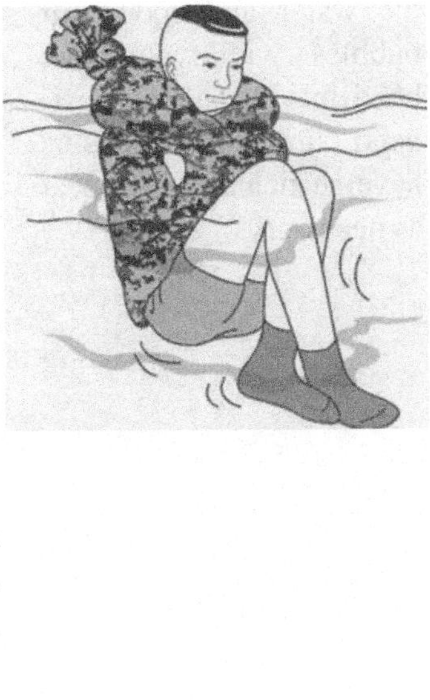

Splash water on the trousers periodically, to prevent the material from drying. Dry material allows air to escape.

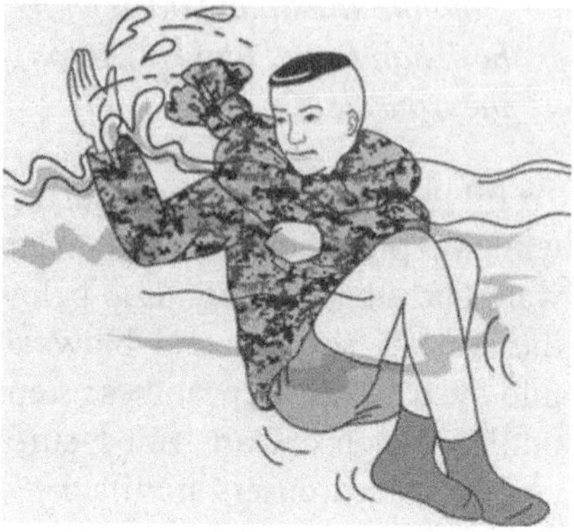

To replenish air in the trousers, you will use a technique referred to as the scooping method. With one hand on the open waistband, extend the trousers in front of you just below the surface of the water and scoop air bubbles with your free hand into the open waist-band until the trousers have sufficient air. Repeat as necessary.

Avoiding Heat Loss in Cold Water

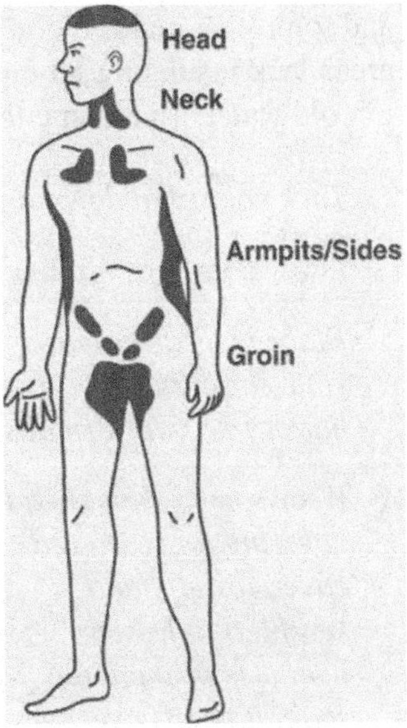

The rate of heat exchange in the water is about 25 times greater than it is in air of the same temperature. When you are immersed in cold water, hypothermia occurs rapidly due to the decreased insulating quality of wet clothing and as a result of water displacing the layer of still air that normally surrounds the body. You also lose about 50 percent of your body heat through your head; therefore, keep your head out of the water. Other areas of high heat loss are the neck, the armpits/sides, and the groin.

In cold water, DO NOT SWIM TO STAY WARM. Swimming, even with a slow and steady stroke, produces a lot of heat that is lost in the water. The heat loss can produce hypothermia that slows body functions and can result in serious injury or death. Remaining motionless conserves body heat three times longer than swimming. SWIM only if you have a flotation device and the shoreline is visible.

Individual Protection From the Cold

If you are equipped with a life preserver, assume the heat escape lessening posture (known as the HELP position) to slow heat loss and to protect major blood vessels near the body's surface. These areas lack insulating fat and are vulnerable to the chilling effects of cold water. To assume the HELP position—

Tuck your chin down tightly to cover your throat.

Draw your legs up in a fetal position to protect the groin.

Place your arms across your chest, tuck your hands into your armpits.

Wear some type of head covering (e.g., stowed cover, towel, handkerchief) to lessen heat loss through the scalp if head covering is available.

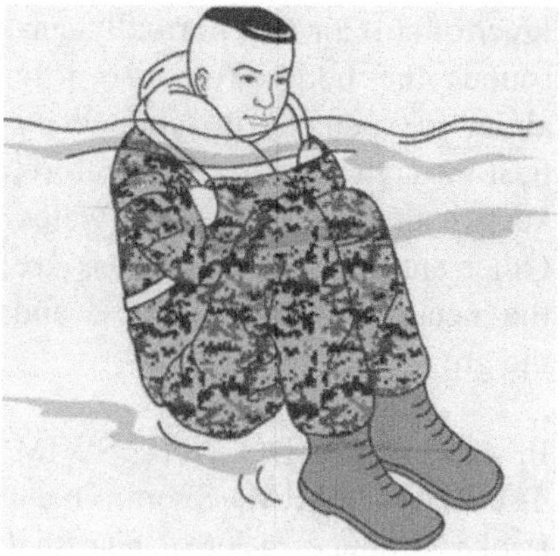

Group Protection From the Cold

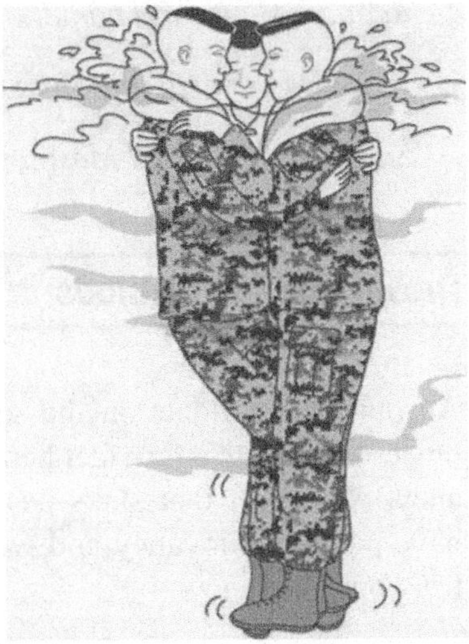

If three or more Marines are in the water and are equipped with life preservers, they should wedge tightly together and lock arms to form a circle known as a huddle position. This position protects vulnerable areas from heat loss.

A casualty who is suffering from the effects of the cold can be placed within the huddle to be surrounded by warmer water. If in the water for a prolonged period, it is recommended that everyone be rotated inside the huddle to maintain or re-warm each person's internal core temperature. If there are more than five Marines, they should make clusters of huddle positions.

Contact with other swimmers provides survival advantages—

Creates a larger target for search and rescue aircraft.

Provides additional warmth in cold water.

Improves morale.

Re-establishes the chain of command.

Reduces shock and panic.

Provides opportunities to administer first aid.

Supports exhausted Marines.

Drownproofing Methods

An object that floats on the surface has positive buoyancy. An object that floats a few feet beneath the surface has neutral buoyancy. An object that sinks has negative buoyancy. Most people have positive buoyancy and will float at the water's surface. To test your buoyancy—

Stand in water that is at your chest level or deeper.

Take a full breath.

Bend forward slowly.

Relax and wait.

If you have positive buoyancy, you will slowly rise to the surface. If you have neutral buoyancy, you will float a few feet beneath the surface. If you have negative buoyancy, you will sink. Regardless of whether you naturally float or sink, you can remain at the surface for extended periods without a life preserver if you exercise the appropriate drownproofing methods or survival strokes, which are based on your buoyancy.

Drownproofing methods consist of the T-method and the sweep. The breast stroke, side stroke, and elementary backstroke are the most common survival strokes. With any drownproofing method or survival stroke, remember the acronym **SAFE**:

SLOW EASY MOVEMENTS
Move slowly to conserve energy and minimize heat loss.

APPLY NATURAL BUOYANCY
Use natural buoyancy to support the body.

FULL LUNG INFLATION
Fill the lungs with each breath. Do not hold air in the cheeks.

EXTREME RELAXATION
Tight muscles are denser than relaxed ones and do not float as well.

Crawl Stroke

The crawl stroke, sometimes called the front crawl or free style, is the fastest stroke. To execute the crawl stroke—

Body position
Lie horizontal, on your stomach, in the water.

Look forward and downward at a 45 degree angle with the waterline between your eyebrows and hairline. Your head position is important as it assists in cutting a path through the

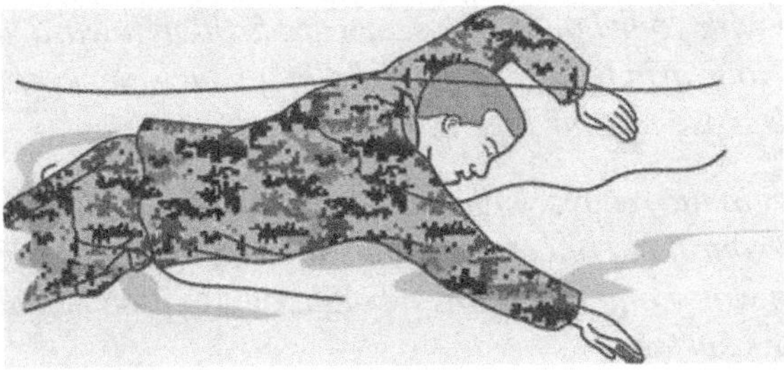

water. If your head is too high in the water, your lower body sinks significantly, making your stroke less efficient. If your head is too low in the water, water washes over your shoulders and neck, causing unnecessary drag.

Arm action

Arm action occurs in three phases: catch, propulsion, and recovery.

Fully extend one arm forward of your body, this positions your hand to catch the water in preparation for the propulsion phase.

To catch the water, bend your wrist (with your palm pointing outboard) and make an "S" shape (or inverted "S" shape) with your hand, ensuring that your hand does not cross the center of your body. Your left hand makes an "S" shape and your right hand makes an inverted "S" shape with your hand finishing at shoulder level.

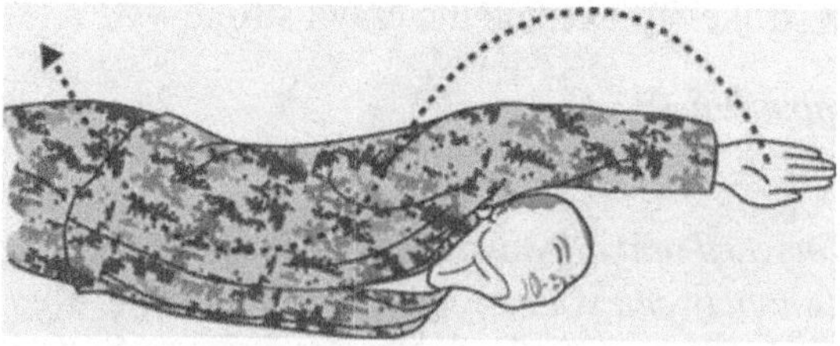

Push with your hands in a rearward fashion toward your feet until your arm is fully extended along your side, keeping your hands close to your body.

To begin the recovery phase, bend your arm at the elbow and raise your hand out of the water. Your hand breaks the surface of the water and maintains a height of 2 to 3 inches above the water's surface.

With your hand and arm moving in a forward manner, bring your hand past your head until your arm is about three-quarters of the way extended. At this point, turn your hand with your palm outboard so as to allow your thumb and fore-finger to enter the water first.

Once your hand enters the water, continue to push your arm forward until it is fully extended.

You are now prepared to catch the water again. These steps are performed in an alternating pattern: when one arm is catching and propelling, the other arm is recovering.

Leg action
Use a flutter kick to create the leg action for the crawl stroke. This kick is used for both propulsion and keeping the lower body horizontal with the water's surface. The flutter kick is an alternating leg action: one leg is kicking in a downward motion while the other leg is recovering to the surface to prepare for the next kick. The size of the kick ranges from 12 to 15 inches and depends on your height.

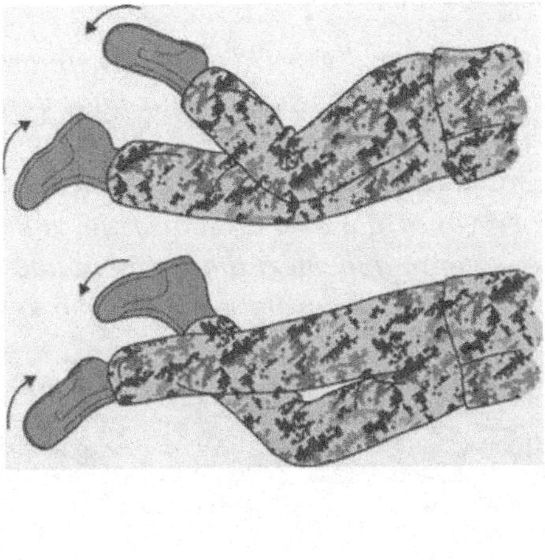

Maintain your legs in a semi-rigid manner.

Generate power for the kick from the hips.

Keep your feet loose at the ankles, they trail behind your legs and act as "flippers."

Execute the propulsion phase of this kick with a downward thrust of your leg.

Execute the recovery phase by pushing your leg back to the surface. This phase is complete when your foot reaches the surface.

Breathing
Breathe during the recovery phase of the arm action.

Roll your body and rotate your head to the side of your body where the arm recovery is occurring, this rolls the water away from your mouth. Keep your chin pushed back toward your shoulder.

Exhale while your face is still submerged and inhale when your face breaks the surface.

Breathe either bilaterally or rhythmically. To breathe rhythmically, breathe on the same side of your body every time your arm cycle occurs. Bilateral breathing requires you to breathe every one and a half arm cycles. To breathe bilaterally, breathe when your right arm is recovering, your face goes back in the water and the next time you breathe is when your left arm is starting its recovery three arm strokes later.

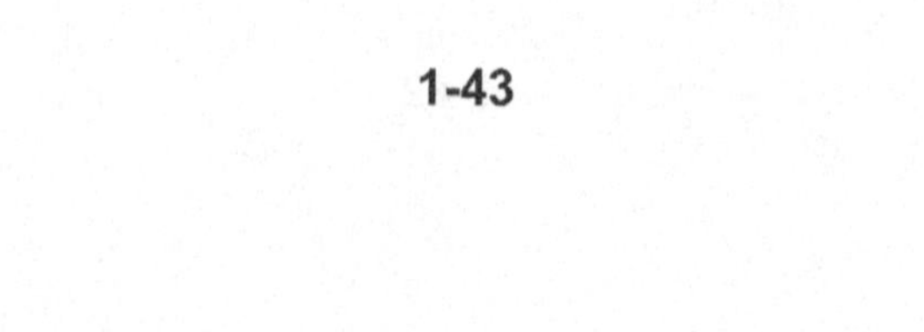

Bilateral breathing is the preferred method for breathing; it prevents the chance of hyperventilation and allows your body to maintain a lateral position in the direction you are swimming.

Coordination
This stroke uses constant arm and leg action.

T-Method

The T-method is a basic drownproofing method. This is the best survival technique if you have negative buoyancy. To execute the T-method—

With your face out of the water, take a deep breath and submerge your face in the water while holding your breath.

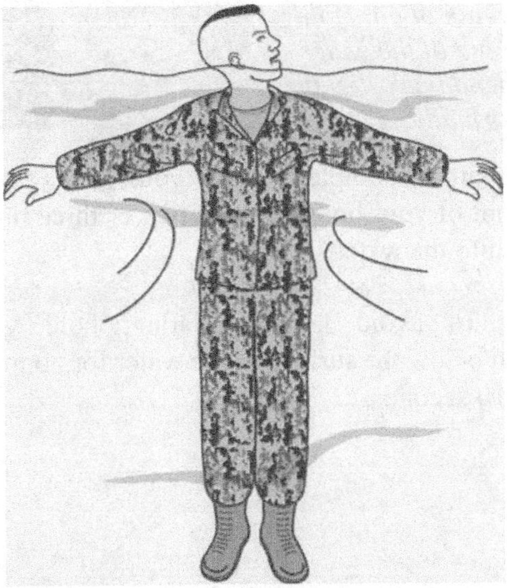

Float with your body in a horizontal position, arms extended from your side, and legs extended and joined.

Move your hands up to your armpits by tracing an imaginary line along your ribs.

Extend your arms outward (horizontal) to your sides, your body position resembles the letter "T."

Step out and forward with one leg and point your other leg to the rear, your knees should be slightly bent. Simultaneously, bring your arms down to your sides. Then bring your legs back together. You complete the steps by exhaling most of your air and preparing to surface your face to obtain another breath of air. Hold your head out of the water. Tilt your head back slightly. Breathe normally.

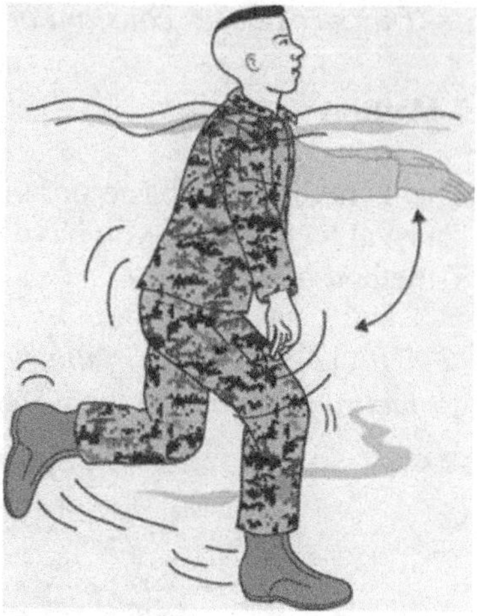

Once your breath is complete, move your hands up and down directly in front of your body. Do this two or three times to slow your descent into the water.

> NOTE: To avoid hyperventilating, hold your breath below the surface of the water for no more than 10 seconds.

The Sweep

The sweep works well if you have slight to excellent positive buoyancy. To execute the sweep—

Float face down in the water: bend 45 degrees at the waist, arms and legs dangling, head hanging down, relax all muscles.

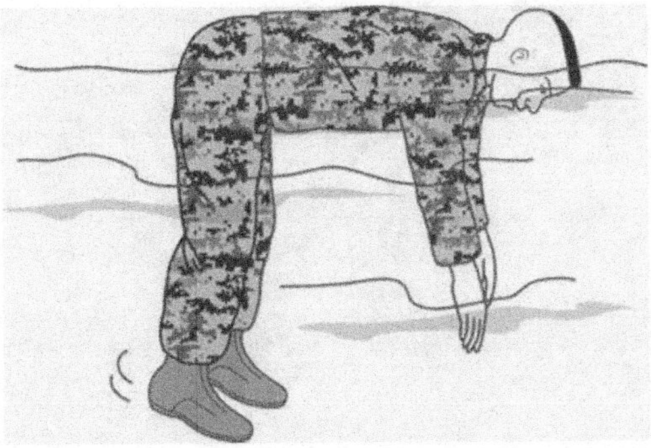

Spread your feet slowly to prepare for a single kick (one leg is forward and one leg is rearward).

Cross your arms in front of your chest, palms outboard with the back of each hand touching the opposite ear.

Exhale prior to raising your head for a breath.

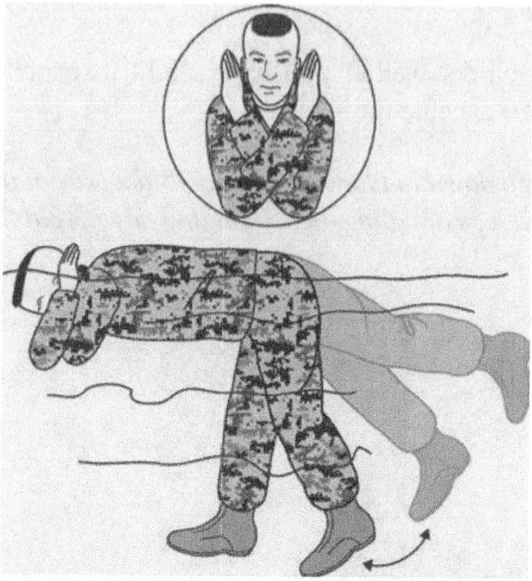

Bring your legs together and sweep your arms down and out until the arms are fully extended out to the sides. This raises your face above the water and allows you to catch a breath of air.

With your air supply replenished, return your face to the water and relax while sweeping your arms in a downward motion in front of your body to prevent/slow your descent. Do not hold your breath for more than 10 seconds. More than 10 seconds enhances your chances of shallow water black out and subsequent drowning.

Breast Stroke

Use this stroke to swim underwater, through oil or debris, and in rough seas. If you are a good swimmer and not wearing combat gear, the breast stroke is the best stroke for long-range swimming because it provides good visibility and allows you to conserve your energy and maintain a reasonable speed.

Body position

Lie prone in the water. Swim with your trunk and legs projecting back and down at an angle of 20 to 30 degrees.

Extend arms out in front (hands together [side by side]), and extend legs behind (toes pointed) to prevent drag.

Face downward, looking forward at a 45 degree angle to break the water and to prevent water from washing into the collar area causing drag.

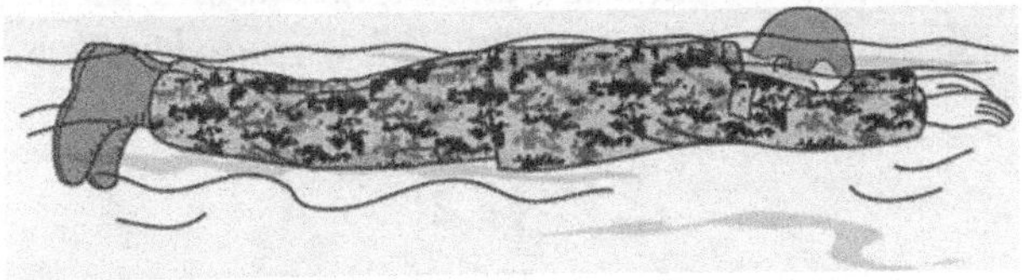

This is known as a glide.

Arm action

Turn your palms outward and bend your arms slightly.

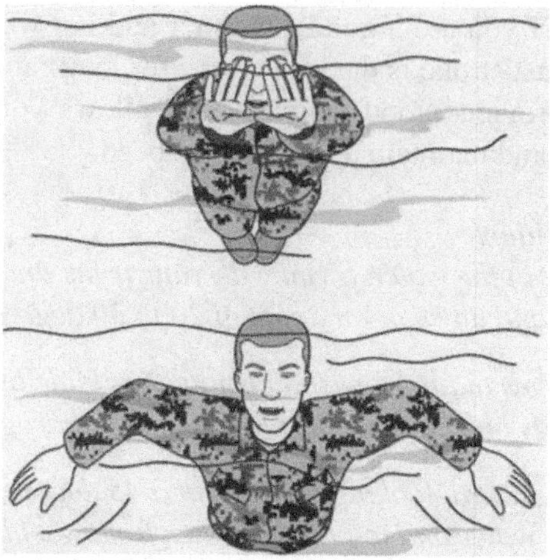

Sweep your arms sideward and slightly downward until your hands are opposite and slightly below your shoulders.

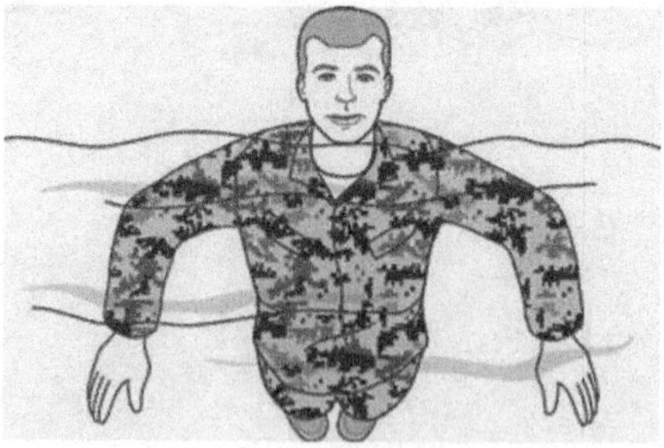

Rotate your head up, breathe once your mouth breaks the surface.

Bring your hands and arms up along your chest and thrust them forward until they are extended and ready to execute the next arm pull.

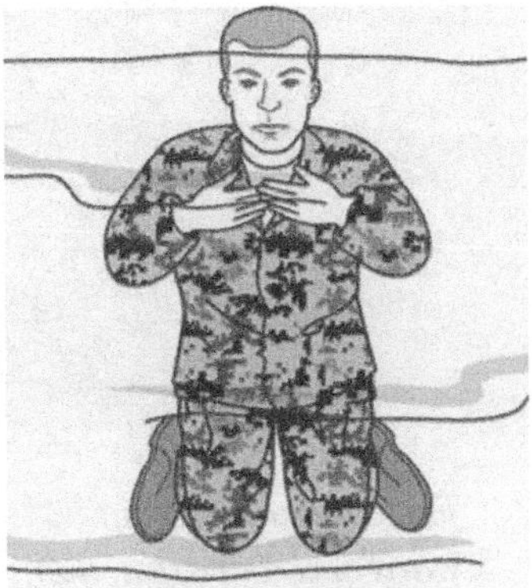

As the arms start their recovery into the glide, the head should rotate forward, resubmerging the face.

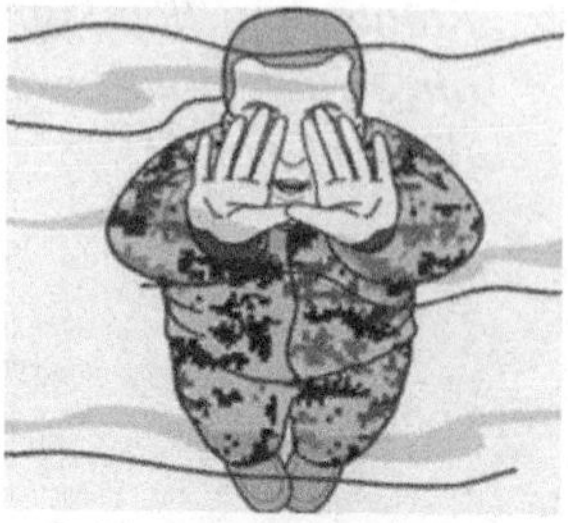

Leg action

Draw your heels toward your buttocks, establish a 45 degree bend in the knees.

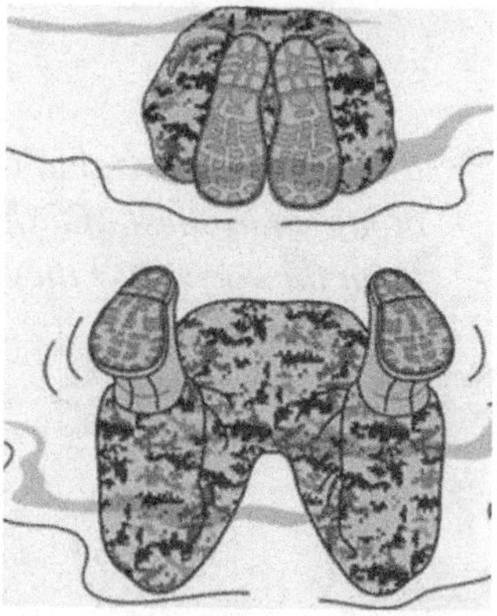

Thrust your legs outward and rearward, then squeeze them together. The whipping action of the feet aids forward propulsion.

This is known as the breast stroke kick.

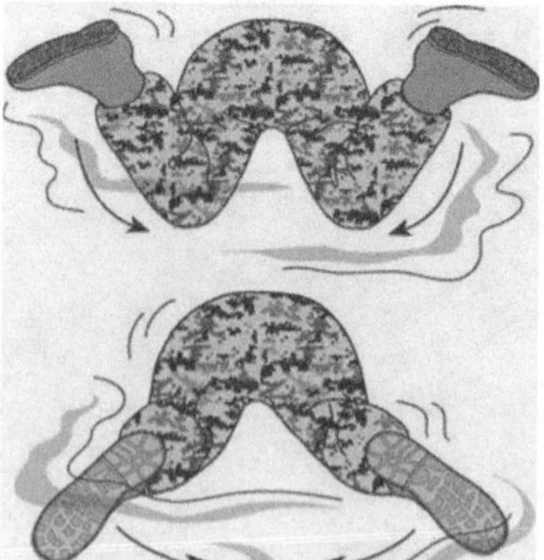

Breathing
Inhale during the arm pull and exhale through your mouth and nose during the finish of the breast stroke kick and glide.

Coordination
The stroke movement is in three counts:

Begin your arm pull. Near the finish of the pull, flex your knees and bring your heels toward your buttocks. The arm pull counteracts the resistance created by the knees.

As the arm pull is completed, thrust your hands forward, kick your legs outward and rearward, and squeeze them together.

Glide through the water for approximately 1 to 3 seconds or until your forward momentum decreases, then begin the next stroke.

Side Stroke

The side stroke is a survival stroke because you use both arms for buoyancy, with each arm creating a slight propulsion. The majority of your body's propulsion comes from your kick. To execute the side stroke—

Body position
Lie on your side with your lead (bottom) arm extended beyond (with a slight bend in your elbow) your head and in line with your body. Palm is down and your hand is submerged 6 to 8 inches.

Extend your trail (top) arm down the length of your body over your thigh.

Keep your legs straight and together, toes pointed rearward.

Keep your face out of the water, this allows for free breathing.

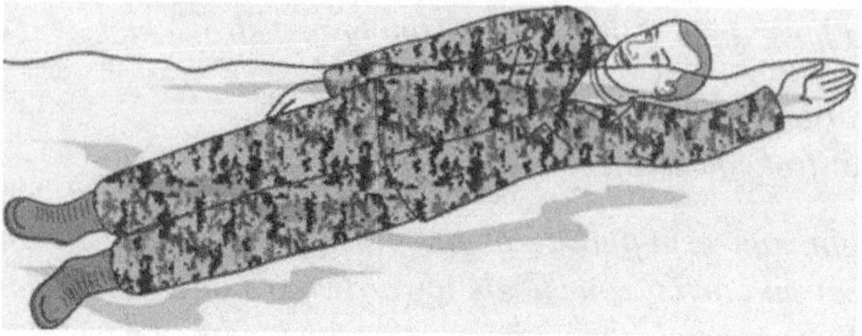

This is known as the glide.

Arm action
With your lead arm, pull your arm downward, while flexing at the elbow, until it is straight down from your shoulder.

Rotate your shoulder and pull your elbow into your side. This should put your lead hand at shoulder level. At the same time, turn your palm toward your face and thrust forward to your original, extended position.

Draw your right hand upward in front of your chest to shoulder level. Rotate your palm toward your feet, then push it downward in front of your body toward your feet to catch the water.

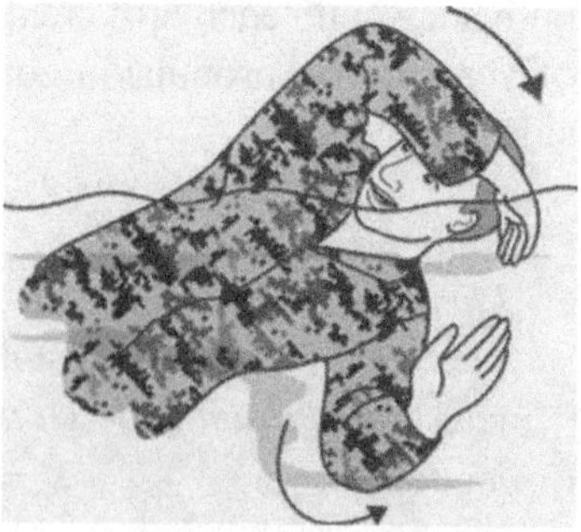

Push your trail hand backward to its original position on top of your thigh. (Your trail hand starts forward and meets your lead hand at your chest/shoulder.)

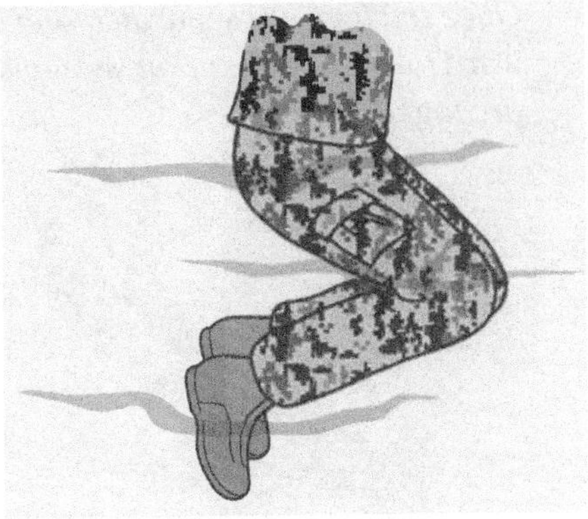

Leg action

To perform the scissor kick, the top leg always goes forward and the bottom leg always goes rearward. From the extended position, draw—or recover—your feet toward your buttocks until your legs are bent at a 45 degree angle at the knees and the hips are flexed at a 45 degree angle with the thighs.

Once the legs have completed their recovery and while maintaining a 45 degree bend in the knees, extend the legs fully into a "V" shape in order to catch the water for the propulsion phase.

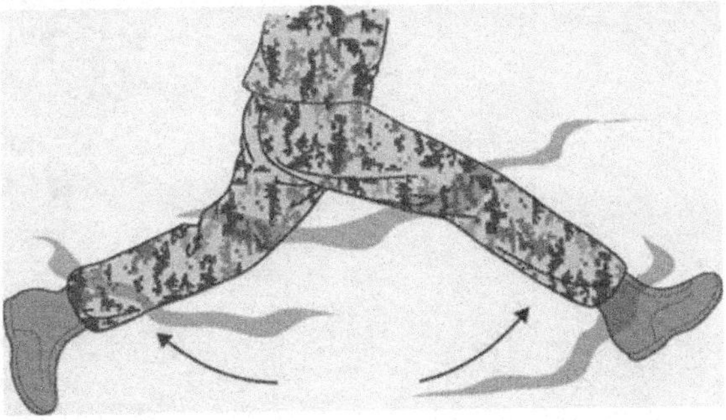

Once the legs are separated and extended forward and rearward to the "V" position, sweep the legs together until the feet are together.

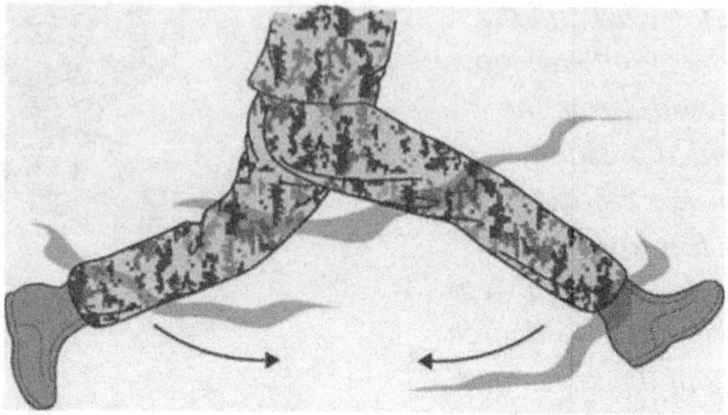

You are now in the glide position.

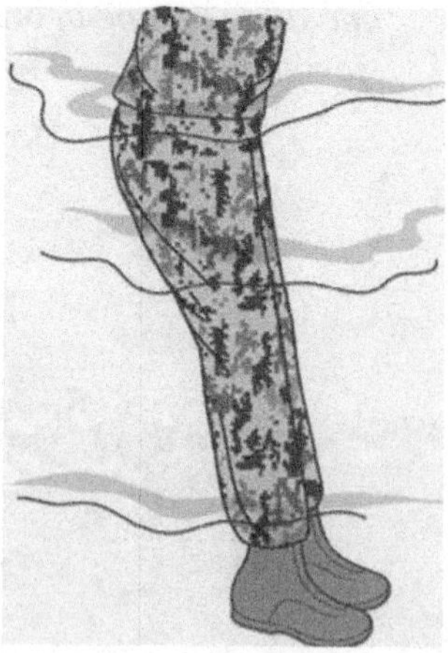

Breathing

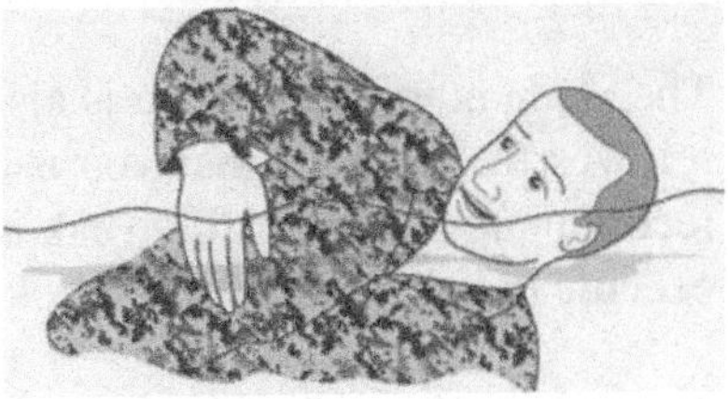

As long as your face remains clear of the water, it is a free breathing stroke. However, it is recommended that you exhale then quickly inhale when the legs are sweeping back together in the scissor kick. This is when the body reaches its highest point in the water, thus clearing the face completely from the surface of the water making it the optimum time to breathe.

Coordination

Begin the stroke with the downward pull of your lead arm. At the same time, bring your trail arm upward and draw your knees up to begin the kick. Let the thrust of the lead arm, push of your trail arm, and the kick of your legs coincide in order to finish the glide position. Glide through the water for approximately 1 to 3 seconds or until your forward momentum decreases, then begin the next stroke.

Elementary Backstroke

The elementary backstroke is also an excellent survival stroke. It relieves the muscles that you use for other strokes, and it is the recommended stroke for weak swimmers or nonswimmers. To execute the elementary backstroke—

Body position
Start on your back.

Face up, chest up, and hips up, keeping an arch in your lower back with arms pressed to your sides and your legs extended and joined to prevent drag.

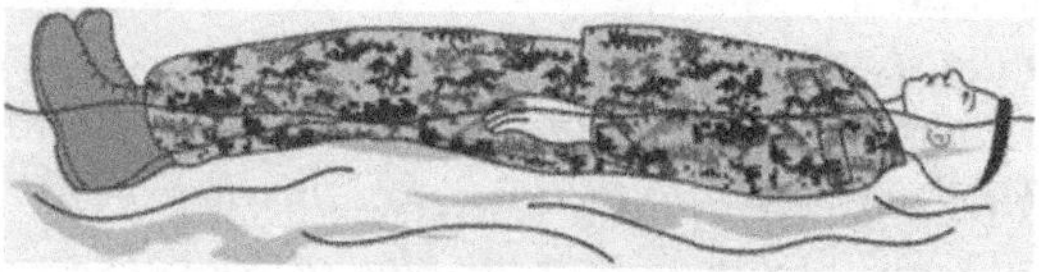

Arm action
Trace your hands up your sides to an area near your armpits then extend your arms out to the sides to form the letter "T" (palms facing feet), locking out the elbows.

> *NOTE:* Don't raise your arms above your head. This creates drag, changes your body position, and submerges the head.

Slap your palms to your thighs using a strong sweeping motion.

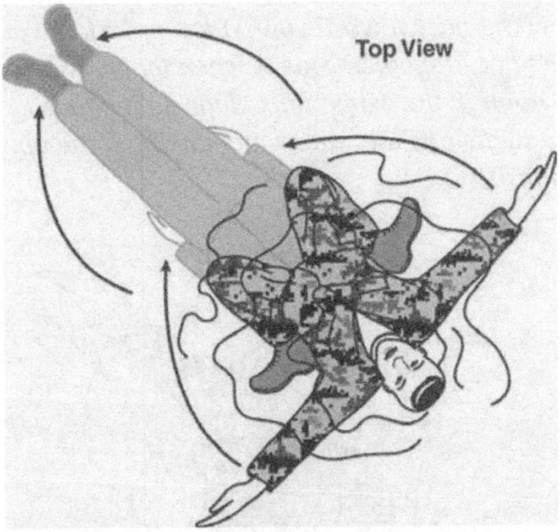

Leg action
Bend both legs at the knee (90 degree angle) slightly separating your knees and drawing your heels downward to a point under and outside your knees. The knees are spread as wide as the hips or slightly wider depending on the body type of the swimmer.

Circle around in a whipping action, ending with legs in a glide position.

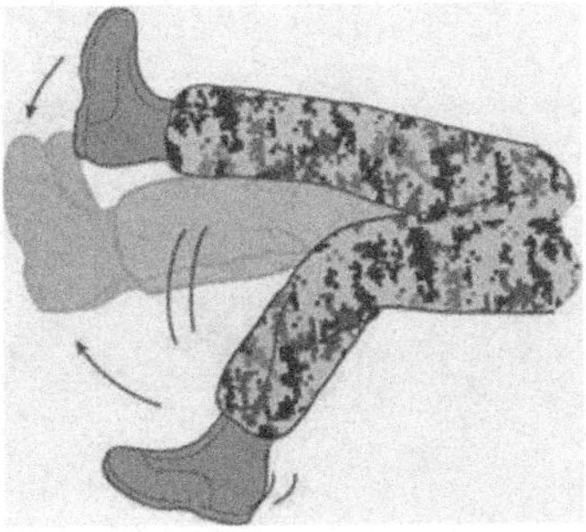

Breathing

Breathe anytime during this stroke. However, it is recommended that you exhale then quickly inhale when your arms are sweeping back toward your sides and while your legs are sweeping back together. This is when the body reaches its highest point in the water, thus clearing the face completely from the surface of the water and making it the optimum time to breathe.

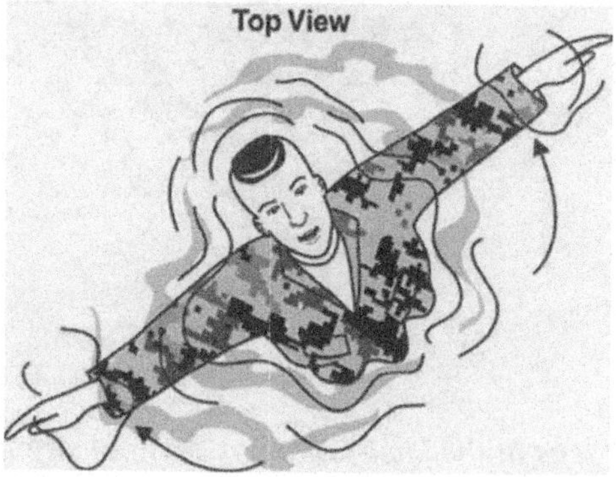

Coordination

The stroke movement occurs in three counts (recovery, catch, power).

Begin the arm pull (recovery).

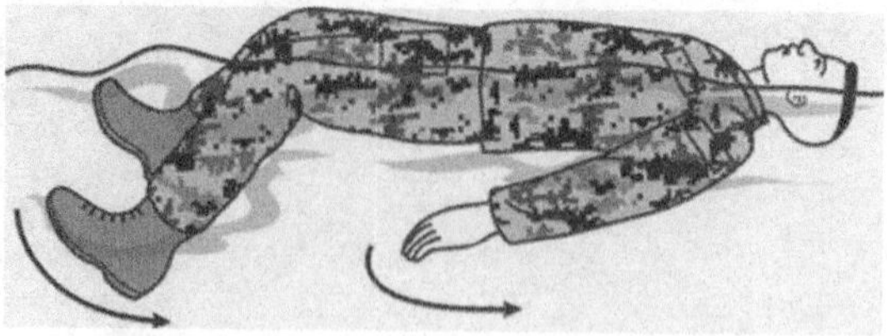

Near the finish of the pull, flex your knees to a 90 degree angle. The arm pull counteracts the resistance created by the knees.

Kick out your legs, and squeeze them together as the arm pull is completed (catch, power).

Glide through the water for 1 to 3 seconds or until your forward momentum decreases, then begin the next stroke as your momentum slows.

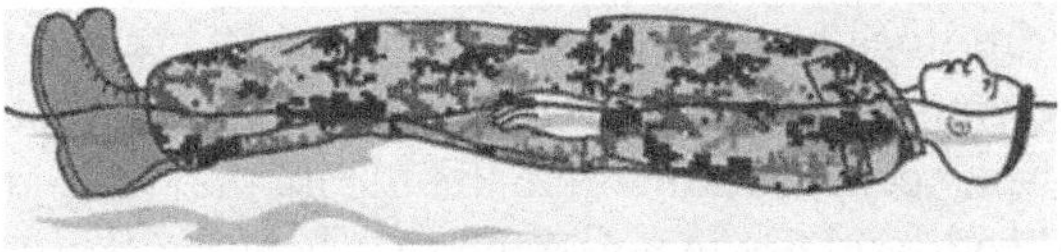

Chapter 2
Water Rescues

A drowning victim can panic and react with unexpected violence and can seize and inadvertently drown a rescuer. Therefore, if possible, a water rescue should be executed from a distance. Reaching, wading, or throwing methods are used in a distance rescue.

If a victim is too far away to use these methods, a swimming rescue may be necessary. Remove all combat gear, when possible, before entering the water. Swim within 2 to 6 yards of the victim to maintain a margin of safety, this allows you to reassess the situation and reassure the victim. If the victim is unconscious, use the wrist tow method or cross-chest carry method to pull the person to safety. If the victim is struggling, use a rear approach and then execute either a single armpit level off or a double armpit level off before towing the victim to safety.

If the victim does begin to overpower you, there are techniques that allow you to defend yourself without having to abandon the rescue. These techniques include the block, the wrist-grip escape, the front head-hold escape, and the rear head-hold escape. These techniques allow you to separate yourself from the victim, reassess the situation, and then attempt the rescue again.

> *Note:* In this chapter's illustrations, the rescuer is shown without a helmet.

Reaching Rescue Techniques

Reach

From a safe position at the water's edge, reach out to the victim. Talk constantly to calm the victim. Retain partial contact with land or some solid support structure (e.g., pier, bridge). If the victim is close but still beyond reach, extend an object (e.g., stick; pack; rifle with magazine removed, chamber empty, and muzzle pointing toward victim) that the victim can grasp. Pull the victim slowly to safety. Once the victim is close to shore, remove the victim from the water.

Note: You can also extend a foot to the victim if you can retain a secure grip on a solid support structure.

Reach From a Deck

Reach from a deck rescue can be executed by either a swimmer or a nonswimmer, and it can be used on an active or a passive victim. Be sure to reassure the victim during the rescue. To execute reach from a deck—

Lie prone on the deck with your body firmly anchored. To anchor your body, lie flat, spread legs apart, and extend one arm behind you with your palm down and on the deck.

Keep as much of your weight on the deck as possible and extend your free hand to the victim.

Grasp the victim's wrist from above, your thumb and index finger are facing you. The victim should never be allowed to grab you and put your life at risk. Therefore, never reach across the victim to grab his wrist, always grab the wrist that is the closest to you.

Keep your arm straight and locked out and pull the victim to the side of the deck. The victim should never be allowed to grab you and put your life at risk. Therefore, never pull the victim into you, always pull the victim into the side.

Arm Extension

An arm extension rescue can be used if you cannot reach the victim using the reach from a deck rescue technique and you must enter the water. An arm extension rescue technique is used for a victim who is either active or passive. Once you determine that a reach from a deck rescue technique is not viable—

Reassure the victim and quickly ease into the water while holding onto the deck with one hand.

Grasp the victim's wrist from above, your thumb and index finger are facing you. The victim should never be allowed to grab you and put your life at risk. Therefore, never reach across the victim to grab his wrist, always grab the wrist that is the closest to you.

Keep your arm straight and locked out and pull the victim to your side. The victim should never be allowed to grab you and put your life at risk. Therefore, never pull the victim into you, always pull the victim into the side.

Leg Extension

If the victim is beyond the reach of your arm, ensure that you have a firm grip on the deck, extend a leg to the victim and allow him to grab it, slowly bring the victim in closer until you can grab the victim's wrist that is holding onto your leg. Once you have a firm grasp on the victim's wrist, use the steps in the arm extension to pull the victim to safety. The leg extension rescue can only be used to rescue an active victim because the victim must be able to grab the rescuer's leg.

Wading Assist

Do not wade into water that is deeper than your chest. Talk constantly to calm the victim. If possible, do not touch the victim directly. Extend an object (e.g., stick; pack; rifle with magazine removed, chamber empty, and muzzle pointing toward victim) that the victim can grasp. Once the victim grasps the object, pull the victim slowly to safety.

Throw

If the victim is not within reach, use an expedient line to throw a lifesaving device to the victim. A lifesaving device can be any weighted item that floats (e.g., a canteen that is one fourth full of water). The lifesaving device must be secured to the end of the rope so that the rope will feed out from its coil when tossed. Talk constantly to calm the victim. Once the victim grasps the line or the lifesaving device, pull the victim toward you at a steady pace that keeps the victim's head above the water's surface. DO NOT pull so strongly that you break the victim's grip on the line. To prepare and use an expedient line and lifesaving device—

Tie a bowline at one end of the rope. (The appendix illustrates various types of knots.)

Unfasten the lid of a canteen.

Place the bowline around the neck of the canteen.

Refasten the lid so the canteen hangs from the bowline loop.

Place one end of the rope under the ball of your forward foot to secure it, either tie a knot at the end of the rope or tie an object at the end of the rope to create a block. Stand with your weight on the end of the rope. DO NOT tie the rope around your ankle.

Coil 20 to 30 yards of the rope, and hold it in a nonthrowing hand.

Place the canteen in your throwing hand.

Use an underhand throw to pitch the canteen and rope a short distance over the victim's head. Keep your nonthrowing hand open so the coil can unfold freely. The rope should trail across the victim's outstretched arms.

Retrieve the rope if the throw is inaccurate or the victim fails to grasp it. Recoil the rope as it is retrieved.

Divide the coil and throw again.

Lifesaving Approaches

Properly approaching the victim is as important as any other aspect of a rescue. Knowing the appropriate and safest method needed to approach a victim is critical in ensuring your safety and the victim's survival. Determining the victim's physical state (distressed swimmer, victim that is active and drowning, or victim that is passive and drowning) is crucial and will determine which type of approach you will execute.

Front Surface Approach

The front surface approach is typically performed when the victim is passive. You must remember that approaching the victim from the front is dangerous because a distressed swimmer or victim that is active can lunge toward you.

Rear Approach

If the victim is active, you must remember that his response can change rapidly. Therefore, approaching him from the front could be extremely dangerous. The victim could easily grab you and take you under the water during a state of panic. Therefore, use a rear approach when possible. When approaching a victim from the rear, reassure the victim until you get to within 2 to 6 yards of the victim; at that point, stop talking to the victim in order to maintain an element of surprise.

Approach Strokes

Before approaching a victim, you must evaluate the victim:

- Is the victim a distressed swimmer?
- Is the victim active and drowning?
- Is the victim passive and drowning?

Select your approach stroke based on your evaluation. Remember, the safety of the rescuer is paramount. Do not endanger yourself in an attempt to reach a drowning victim.

Crawl Stroke Approach Stroke

The crawl stroke (free style) is the fastest approach stroke. This stroke is used when the victim is passive and/or unconscious in the water. That is, the victim is face down, submerged or near the surface of the water, is not breathing, or is not moving. If any of these conditions exist, there is an obvious need to reach the victim as quickly as possible.

Execution of the crawl stroke approach stroke is the same as the crawl stroke (see pages 1-40 through 1-44), except that your head remains above the water's surface to allow for free breathing and maintaining eye contact with the victim.

Maintaining eye contact is critical. If the victim becomes submerged, you will have a better chance of locating him if you know his last location. Your distance from the victim will determine if your head remains out of the water the entire time you are executing the stroke. If you are 54 yards or less from the victim, you should maintain eye contact by keeping your head raised until you reach the victim. If you must swim more than 54 yards to reach the victim, make eye contact with the victim, place your face back in the water for four or five strokes, then raise your head again to regain eye contact with the victim (repeat these steps until you reach the victim).

Body position
The position of your body is horizontal to diagonal and influenced by the position of the head. This allows you to maintain eye contact with the victim.

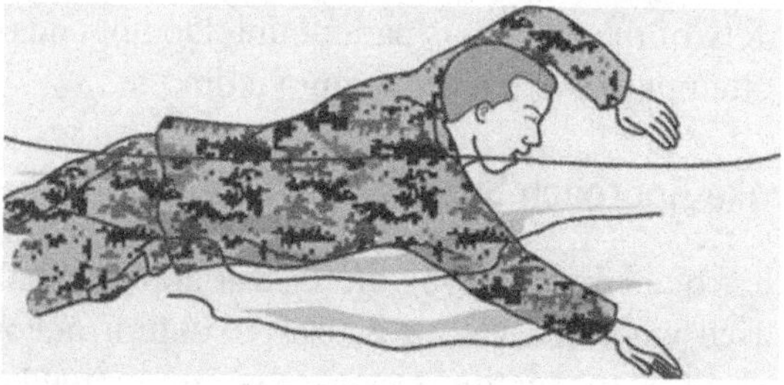

Arm action
While one arm is forward of your head to catch the water for the propulsion phase, extend your other arm alongside your body in what is known as the recovery phase. The recovery phase ends when your arm is extended forward of your head to begin the propulsion phase. In the propulsion phase, your arms are approximately shoulder-width apart and the pulling action is slightly wider and deeper to compensate for your raised head.

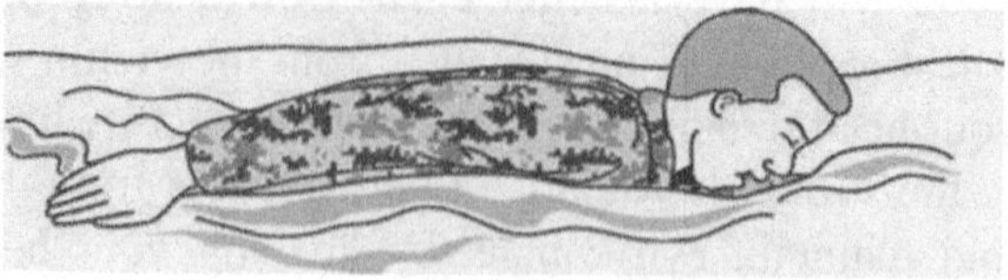

Leg action
Kick one leg in a downward motion while your other leg is recovering to the surface to prepare for the next kick (this is known as a flutter kick). This kick is used for propulsion and maintaining the lower body as horizontal to the water's surface as possible. When your head is raised, your knees are slightly bent in order to keep your feet near the surface.

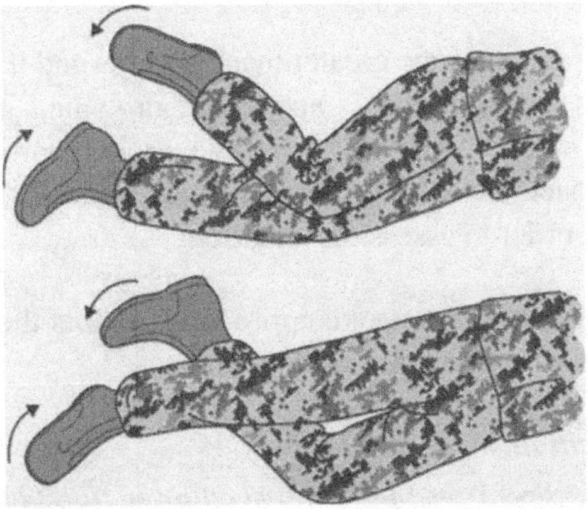

Breathing
If your face is above the water, breathe as needed (free breathing).

If traveling in excess of 54 yards to reach the victim, rotate your head to the right or left during the recovery phase and inhale when your face is out of the water or wait until your head is out of the water and facing forward while regaining sight of the victim.

Coordination
This stroke uses constant arm and leg action.

Breast Stroke Approach Stroke

The breast stroke approach stroke is used when a swimmer is distressed or a victim is active and drowning. This stroke is also used when the victim has a suspected spinal injury because it minimizes the wake created around the victim, minimizes the movement of the victim's head and neck, and helps prevent further injury.

If a victim is at a distance greater than 54 yards and if the swimmer is in distress or the victim is active and drowning in open water (ocean, lake), you should use the crawl stroke approach stroke to rapidly approach the victim, stopping approximately 11 yards from the victim in order to assess the situation.

Execution of the breast stroke approach stroke is the same as the breast stroke with a few minor deviations.

Body position
The position of your body is horizontal to diagonal and influenced by the position of the head. This allows you to maintain eye contact with the victim. It also allows you to communicate with the victim in order reassure him as you approach.

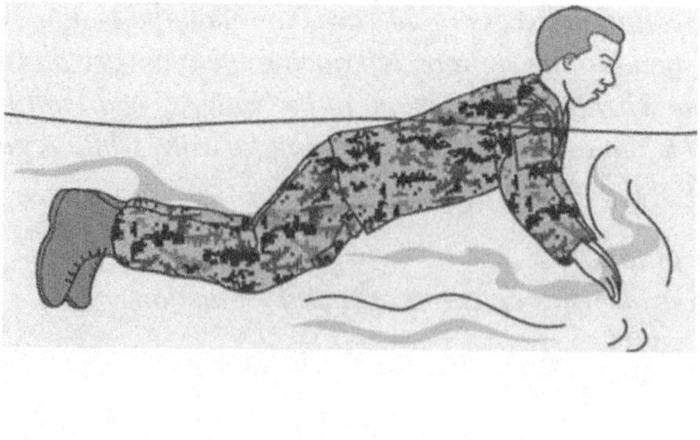

Arm action
Execute the arm action for the breast stroke, except that your arm pull is wider and deeper, which allows your head to remain above the water's surface.

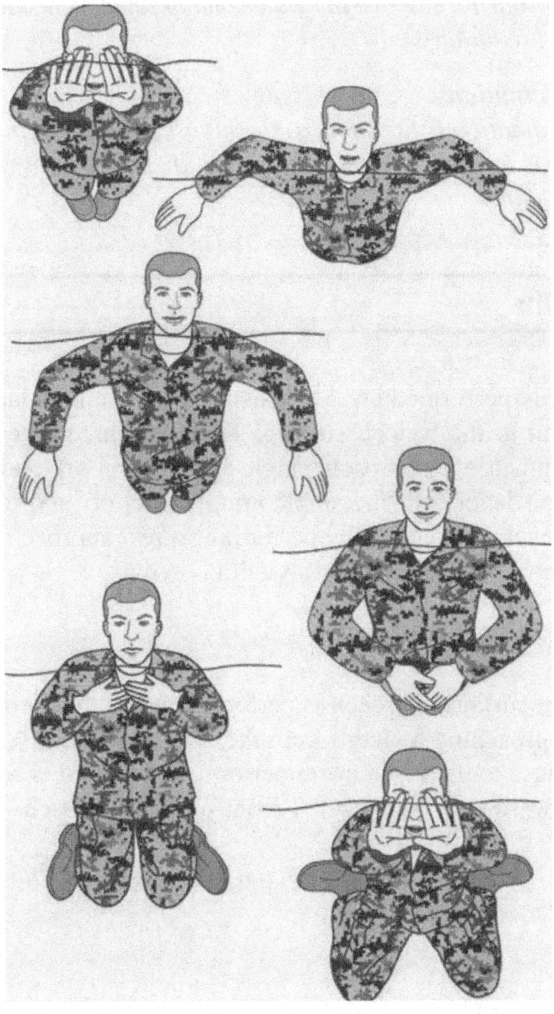

Leg action
Execute the leg action for the breast stroke, except that your legs open wider during the propulsion phase of the stroke.

Breathing
Keep your face above the water and breathe as needed (free breathing).

Coordination
Coordination is the same as the breast stroke, except that there is no glide period. You constantly stroke until you reach the victim.

Level Offs

A properly performed level off positions the victim's face up and horizontal to the water's surface. There are three types of level offs: front surface approach, single armpit level off, and the double armpit level off. The single armpit level off and the double armpit level off are done by performing a rear approach, which is the preferred approach when a victim is active.

Front Surface Approach

The front surface approach is performed when the victim is passive. Approaching a victim from the front may place you in danger because a distressed swimmer or a victim who is active may lunge toward you. To execute a front surface approach—

Stop 2 to 4 yards from the victim to reassess the situation and reassure the victim.

Determine which arm to use in order to rotate the victim into a face-up position (your right hand on the victim's right wrist or your left hand on the victim's left wrist).

Move into position and prepare to grasp the victim.

Turn sideways and move toward the front of victim.

Once in position, reach forward and grab the victim's wrist. Your thumb is on the inside of the victim's wrist, as if you were checking his pulse with your thumb. Your remaining fingers wrap around the victim's wrist.

Lean back immediately and execute a powerful scissor kick or inverted scissor kick and perform short, vigorous pulls with your free arm.

As the victim begins to move forward in the water, kick, pull, and twist outboard on the victim's wrist. The momentum created from the kick and the pulling and twisting action of the victim's wrist will rotate the victim into a face-up and horizontal position in the water.

Extend and lock out your towing arm down the length of your body and execute a wrist tow to move the victim to safety.

Single Armpit Level Off

Approach the victim.

Stop 2 to 6 yards from the victim to reassess the situation, reassure the victim, and maintain your margin of safety.

Approach the victim slowly and grab his armpit with your hand (your right hand to victim's right armpit or your left hand to victim's left armpit). Position yourself sideways to the victim and place your elbow in the center of his back.

Pull with your hand that is in the victim's armpit while pushing with your elbow that is in the victim's back to place the victim horizontal on the water's surface (face up). To assist in placing the victim horizontally, use your free arm to execute short, vigorous pulls and use your legs to execute a scissor kick or inverted scissor kick.

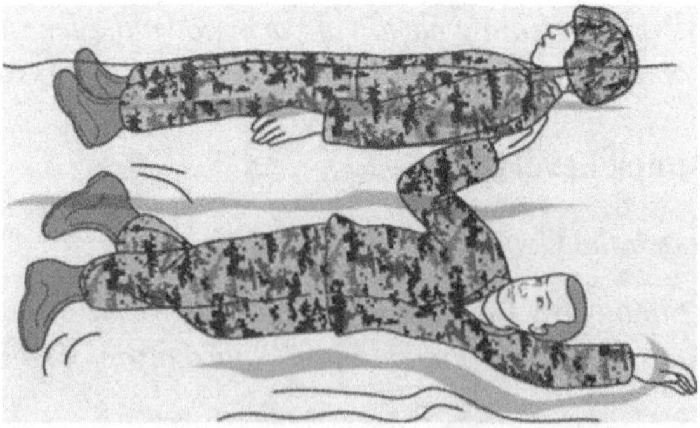

Once the victim is horizontal and on the water's surface, begin forward momentum by extending your towing arm to a fully locked-out position and executing a single armpit tow.

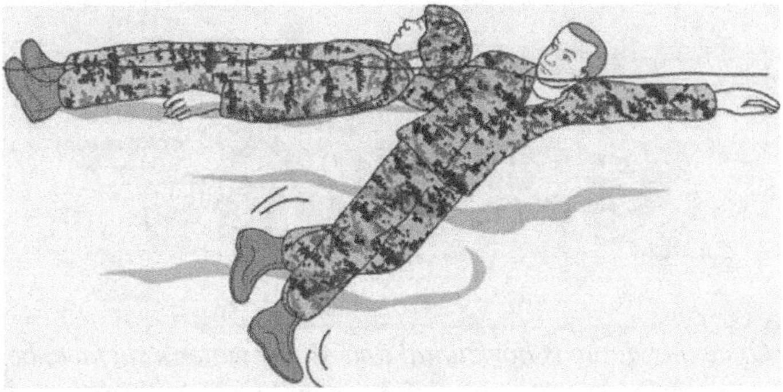

Double Armpit Level Off

Approach the victim.

Stop 2 to 6 yards from the victim to reassess the situation, reassure the victim, and maintain your margin of safety.

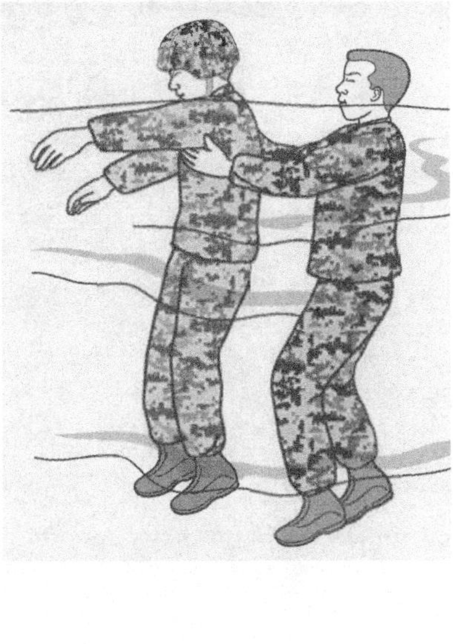

Approach the victim slowly and grab his armpits with both your hands. Place your elbows on his back.

Pull with your hands that are in the victim's armpits while pushing with your elbows that are in the victim's back and use an inverted breast stroke kick to place the victim horizontal on the water's surface (face up).

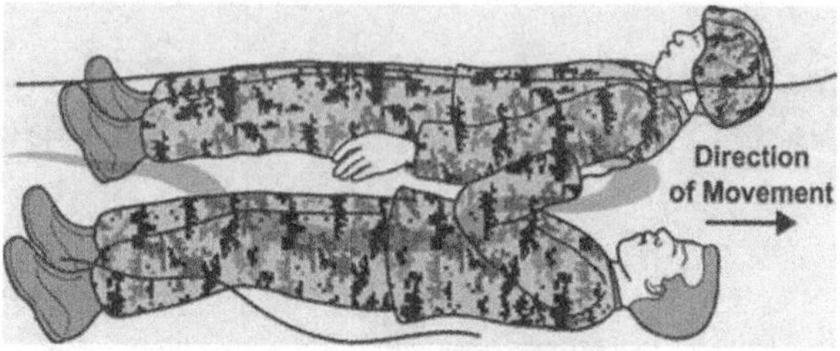

Once the victim is horizontal and on the water's surface, begin rearward momentum by slowly extending your arms into a fully locked-out position and executing a double armpit tow and inverted breast stroke kick.

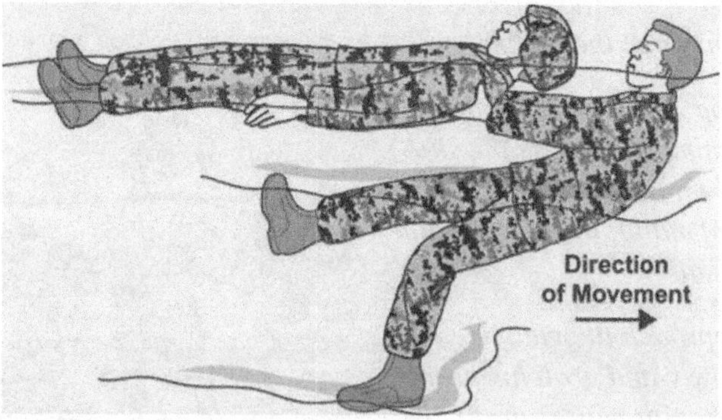

Rescue Techniques

Wrist Tow

Use the wrist tow method to rescue a victim who is floating face down. DO NOT use the wrist tow on a struggling victim. If time allows, remove your helmet and gear before attempting the rescue. Swim toward the victim using a modified breast stroke. Swim within 2 to 6 yards of the victim to maintain a margin of safety, this allows you to reassess the situation and reassure the victim. The following steps show proper front surface approach, wrist tow procedures:

Approach the victim from the front and grasp the underside of the victim's left wrist with your left hand or the right wrist with your right hand. Ensure that your thumb is on the underside of victim's wrist.

Lean back, pulling and kicking strongly to move the victim into a horizontal position.

Twist the victim's wrist to rotate the victim into a face-up position.

Swim toward safety using the lifesaving stroke.

> NOTE: The lifesaving stroke is a modified side stroke wherein the top arm is used to tow or carry a victim to safety and use a scissor kick or an inverted scissor kick is used for propulsion.

Keep a firm grip on the victim's wrist.

Keep your towing arm fully extended and along your side. This keeps the victim in the water column and prevents drag.

Ensure the victim's head does not go under water during the recovery.

Single Armpit Tow

You perform the single armpit tow after a victim has been properly leveled off. The single armpit tow uses the lifesaving stroke. To execute the single armpit tow—

Place one hand (top arm) in the victim's armpit. Your towing arm is straight and along your side.

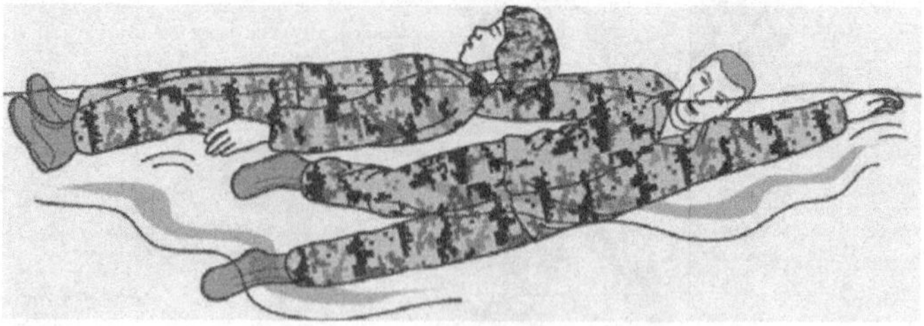

Use your lead arm (bottom arm) to execute short, vigorous pulls.

Execute a scissor or inverted scissor kick in a continuous and vigorous manner.

Use either free breathing or explosive breathing during the tow. During free breathing, keep your head above the water's surface and continuously reassure the victim. During

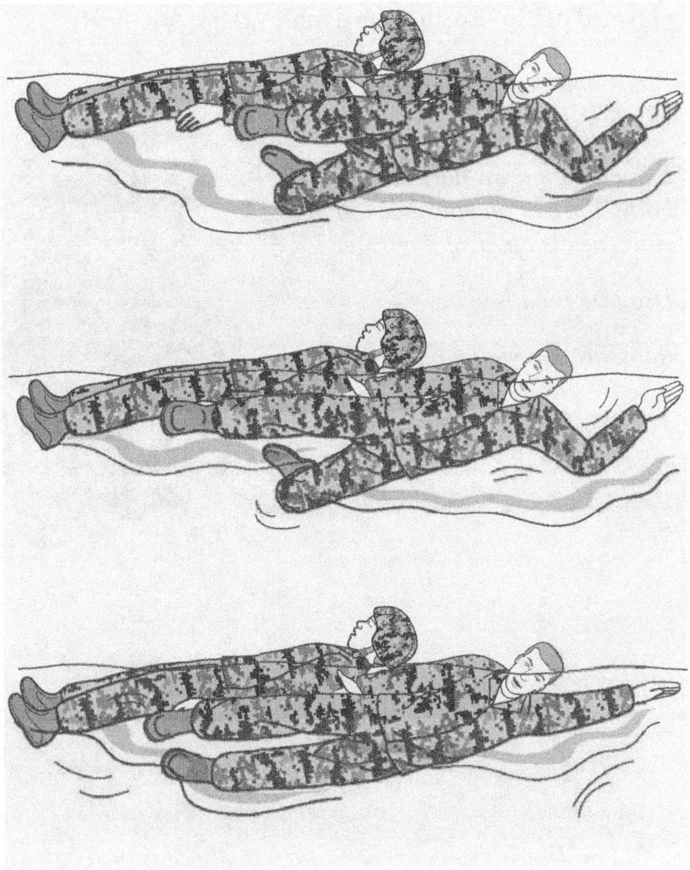

explosive breathing, put your head under the water to plane off your body's angle and reduce drag. Execute approximately two strokes, then lift your head up to breath and reassure the victim. Return your head back underwater for the next few strokes.

Double Armpit Tow

You perform the double armpit tow after a victim has been properly leveled off. To execute the double armpit tow—

Place both hands in the victim's armpits.

Extend both arms fully along your body. You are on your back, and your face is clear of the water.

Use free breathing.

Reassure the victim at all times.

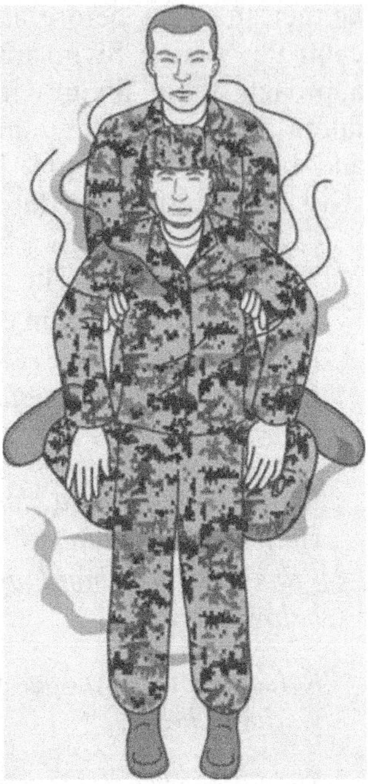

Use the inverted breast stroke kick (it is the only kick that can be used due to your body position) to tow the victim to safety. (The inverted breast stroke kick is the same kick used in the elementary backstroke.) Your kick must be continuous and vigorous in order to keep the victim's face above the water.

Collar Tow

You perform the collar tow after a victim has been properly leveled off. To perform a collar tow using the victim's blouse—

Maintain control of the victim by grasping his armpit with one hand and then with your free hand grasp either his combat gear or his blouse between his shoulder blades. If grasping the blouse, grasp the material with your palm up, then turn your hand over to tighten the material.

Release the victim's armpit once control is established and execute the lifesaving stroke to tow the him to safety.

Cross-Chest Carry

Use the cross-chest carry to carry a victim to safety if the victim is struggling or when moving through heavy surf. Remove your helmet and gear before attempting a rescue. Talk constantly to calm the victim. Swim toward the victim using a breast stroke approach stroke. Swim within 2 to 6 yards of the victim to maintain a margin of safety, this allows you to reassess the situation and reassure the victim. The following steps illustrate proper cross-chest carry procedures:

> **CAUTION:** The cross-chest carry causes fatigue even if you are in excellent physical condition.

Use a level-off technique to place the victim in a horizontal, face-up position.

Retain a grip on the victim with one hand. Reach over the victim with your free hand to encircle the victim's chest. Place your free hand on the victim's opposite rib cage, just below his armpit.

Release your grip once you have a secure hold on the victim's chest.

Swim toward safety using the lifesaving stroke while keeping a firm grip on the victim's chest and your hip on his back.

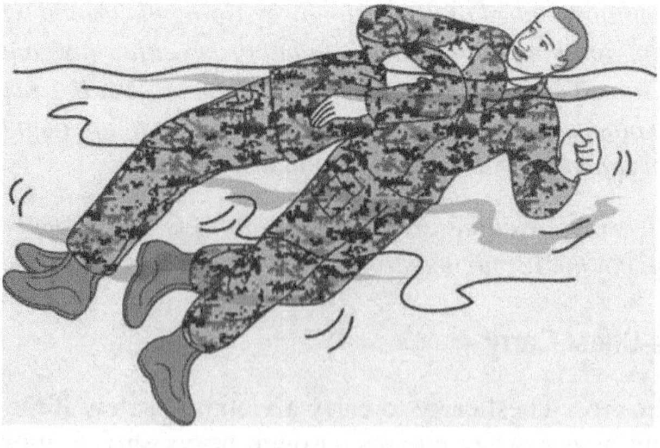

This procedure brings the victim's face and shoulders clear of the water, and typically the victim stops struggling. Sometimes, however, the victim will struggle during the swim to safety. If this happens, either tighten your grip on the victim or defend yourself with one of the techniques discussed on pages 2-25 through 2-30.

Tired Swimmer's Assist

You must maintain a 2 to 6 yards margin of safety from the victim at all times while you are getting into position to perform the tired swimmer's assist. Once behind the victim, you extend one arm, hand straight. Place your straight hand underneath the victim's opposite armpit. Maintain a 45 degree angle away from the victim, your arm is locked out. Assist the victim in propulsion until both of you reach safety.

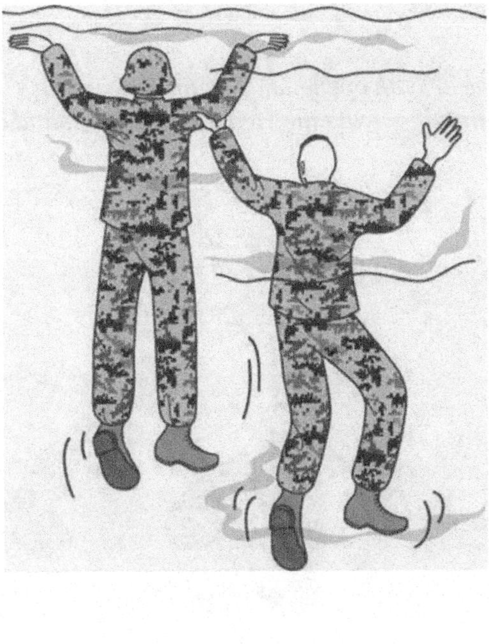

Defense Against a Drowning Victim

DO NOT sacrifice your life in an attempt to save the victim. A struggling drowning victim poses great danger to anyone nearby. Driven by panic, the victim can grab you with great strength in an effort to climb out of the water. This can result in death for both you and the victim. The following techniques can be used to defend against a drowning victim, but the best defense against attack by the victim is to stay out of reach.

Block

The block prevents the victim from grabbing you if you have approached from the front. If the victim lunges toward you, react as follows:

> *Place one or both open hands against the victim's upper chest; being careful to avoid the victim's face, neck, and abdomen.*

Lean backwards and submerge rapidly. Keep your blocking arm(s) extended.

Swim underwater and away from the victim, and return quickly to the surface.

Stop 2 to 6 yards from the victim to reassess the situation.

Determine an appropriate course of action.

Wrist-Grip Escape/Wrist-Grip Escape Alternative

The wrist-grip escape is used when a victim grabs your arm or wrist. Submerge the victim quickly by reaching across with your free hand, pushing down on the victim's shoulder to submerge him, and kicking to propel yourself upward. While keeping the victim submerged with your hand on his shoulder, give three hard jerking pulls with your trapped hand in an attempt to break free from his grasp. Once free, swim clear of the victim and reassess the victim's condition.

If the wrist-grip escape is unsuccessful in freeing your hand, use the wrist-grip escape alternative: use your free hand to grab your trapped fist, rotate thumbs up, apply bone-on-bone contact with the victim's arm, pry your hand out of his grasp, and quickly swim away from the victim.

Front Head-Hold Escape

The front head-hold escape allows you to escape when you are facing a victim who is gripping you around your head and neck. To execute the front-head hold escape—

Take a quick breath and tuck your chin into your shoulder to protect your throat.

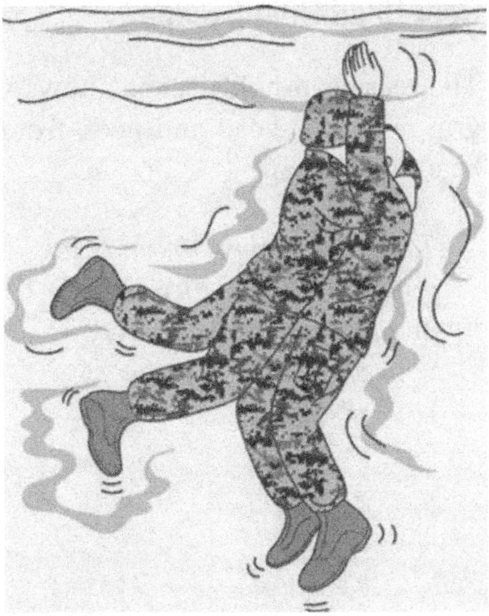

Clap your hands above your head (three times) to submerge instantly. This drags the victim below the water, lifts his arms from around your neck, and, typically, he releases his grasp in order to get back to the surface. If he doesn't release his grasp, apply pressure to the victim's brachial pressure points (which are located inside of the upper arm, above the elbow).

Thrust the victim's arms up and away.

Keep your chin tucked to protect your throat, and swim underwater away from the victim and return quickly to the surface at the ready position.

Stop 2 to 6 yards from the victim to reassess the situation.

Determine an appropriate course of action.

Rear Head-Hold Escape

The rear head-hold escape allows you to escape when a victim is gripping your head and neck from the rear. To execute the rear head-hold escape—

Take a quick breath. and tuck your chin down, turn your head to either side, and raise your shoulders to protect your throat.

Take a strong stroke, clap your hands above your head (three times), and submerge instantly. This drags the victim below the water and, typically, he releases his grasp in order to get back to the surface. If he doesn't release his grasp, apply pressure to the victim's brachial pressure points (which are located inside of the upper arm, above the elbow).

Thrust the victim's arms up and away.

Twist your head and shoulders until free.

Swim underwater away from the victim and return quickly to the surface.

Stop 2 to 6 yards from the victim to assess the situation.

Determine an appropriate course of action.

Administering First Aid/Rescue Breathing

To administer first aid/rescue breathing, remove the unconscious victim from the water when possible and stop any bleeding.

First, open the airway and check for breathing. If the victim is not breathing, give two breaths of air. Check for a pulse. If there is a pulse but the victim is not breathing, continue rescue breathing. If there is no pulse, start cardiopulmonary resuscitation (CPR). For detailed information on administering CPR and rescue breathing, refer to American Red Cross CPR skill cards.

WARNING

> If the victim has no pulse and is not breathing, administer CPR immediately. If the victim does have a pulse but is not breathing, give rescue breathing only. If the victim has a pulse and is breathing, DO NOT give CPR—CPR could prove fatal.

Next, protect any wounds from exposure, and treat for shock.

Chapter 3
Treatment of Casualties and Avoidance of Dangerous Marine Life

To survive in the water, you will face many challenges other than just being able to swim. For example, injuries and/or fatigue can lead to drowning and exposure to the elements can lead to hypothermia or heat injuries and may require medical attention. Finally, your presence in the water may be seen as a threat to or food source for a variety of marine life and you may be attacked. This chapter advises you on how to minimize these risks in order to stay alive.

Drowning

Drowning is suffocation by a liquid. A drowning victim inhales water into the lungs or the throat closes by reflex so that little or no water can enter the windpipe. In either case, a victim can no longer breathe.

Symptoms

One symptom of drowning is that the victim may call for help and has an expression of dread or panic. But typically a victim that is active and drowning may not call for help because he is trying to conserve his air and will not speak.

Another symptom of drowning is when the victim thrashes at the water's surface. If the victim stops or grows calmer, he has likely been overcome by fatigue, hypothermia, or a lack of air. At this stage, the victim usually has 20 to 60 seconds before going under the water's surface.

Treatment

If the victim is not breathing, begin rescue breathing. Place the victim on his back, tilt head back to open airway, pinch the nose, and give two full breaths. If the victim does not inhale during the first two breaths, reposition his head and attempt two more breaths. Check for a pulse. If a pulse is present, but the victim is still not breathing, continue rescue breathing. If a pulse is not present, begin CPR. See MCRP 3-02G, *First Aid*, for rescue breathing and CPR details.

WARNING

If the victim has no pulse and is not breathing, administer CPR immediately. If the victim does have a pulse but is not breathing, give rescue breathing only. If the victim has a pulse and is breathing, DO NOT give CPR—CPR could prove fatal. Continue first aid until medical help arrives.

A victim who is not breathing and has no pulse may appear dead. DO NOT decide that death has occurred. Continue with the prescribed treatment. A corpsman or medical officer should decide whether the victim can or cannot be revived.

Hypothermia

Hypothermia is the abnormal lowering of the body's internal (or core) temperature to 95 degrees or below. It occurs when the body loses heat faster than it produces heat. A hypothermia victim loses the ability to move quickly, becomes mentally sluggish, slips into semiconsciousness, lapses into a coma, and dies when internal body temperatures drop too low.

The chilling effects of cold air, wind, or water can produce hypothermia. Water poses the greatest threat because it transfers heat 25 times faster than air. Depending on the water's temperature, a victim can succumb to hypothermia within a few minutes. The body's sudden contact with cold water can also set off a body reaction known as the mammalian diving reflex. This reflex can greatly increase survival time (especially for women and children) in or under cold water. The mammalian diving reflex shuts off blood circulation, except for the flow between the heart, lungs, and brain. The small amount of oxygen left in the blood and lungs is saved for the body's vital organs. This reflex has allowed people to survive being under cold water for an extended period of time. Therefore, a cold water drowning victim should be treated as if still alive even though the victim is not breathing, has no pulse, and may appear dead. DO NOT decide that death has occurred. Continue with the prescribed treatment. Victims of hypothermia can appear to be dead when they are not. A corpsman or medical officer should decide whether the victim can or cannot be revived.

WARNING

If the victim has no pulse and is not breathing, administer CPR immediately. If the victim does have a pulse but is not breathing, give rescue breathing only. If the victim has a pulse and is breathing, DO NOT give CPR—CPR could prove fatal. Continue first aid until medical help arrives. Check for a pulse for at least 45 seconds.

Symptoms

Once the body's core temperature drops, the victim will show one or more of the following symptoms:

- Violent and uncontrollable shivering as the body tries to warm itself.
- Slow or slurred speech.
- Disorientation or poor coordination.
- Loss of skin color.
- Blue and pinched lips.
- A slowing or stopping of shivering that progresses into a rigid torso and limbs.

Survival Time

A hypothermia victim's survival depends on the water's temperature and the time spent in the water. A small body build cools faster than a large build. Children cool faster than adults.

To increase your chance of survival in the water, utilize the HELP position described on page 1-37. Extra clothing and inactivity (remaining motionless in the water) can also increase your survival time.

Treatment

A hypothermia victim must be warmed to prevent further heat loss; therefore, treatment should begin as soon as possible. Consciousness of the victim determines the treatment that should be pursued. See MCRP 3-02G for specific treatment information. The following treatment procedures are recommended:

- If the victim is conscious, give the victim warm fluids. Give candy or sweetened foods to a victim who is able to eat.
- If the victim is unconscious, place him on his back with his head tilted back to ensure an open airway.
- DO NOT massage the victim. Massage can break blood vessels and create swelling, internal pressure, and blocked blood circulation.
- DO NOT give alcohol to the victim. Alcohol lowers the victim's body temperature.
- Shock is a possibility, treat accordingly.
- Seek medical help immediately.

If you are able to remove the victim from the water, apply the following steps when possible:

- Get the victim into shelter.
- Remove the victim's wet clothing.
- Put the victim in dry clothing.
- Place the victim in a sleeping bag if one is available. It may be necessary to place another Marine in the sleeping bag with the victim.
- Place as much insulation as possible between the victim and the ground.
- Use hot water bottles, electric blankets, or blankets heated in an oven or by a campfire to warm the victim's neck, groin, and the sides of the chest.

CAUTION: DO NOT apply heat to extremities.

Heat-Related Injuries

Heat-related injuries include heat cramps, heat exhaustion, and/or heat stroke. They can occur when a Marine—

- Is exposed to extreme heat, such as from the sun or a combination of high air temperatures and water temperatures.
- Does not wear proper clothing or gets overexposed to the sun.
- Becomes dehydrated.

With heat cramps, muscles may cramp in the arms, legs, and/or stomach. The victim may sweat excessively. To treat, create improvised shade for the victim and have him drink water.

With heat exhaustion, the victim sweats heavily; presents pale, moist, cool skin; and complains of a headache, weakness, dizziness, and/or loss of appetite. A victim may also experience heat cramps, nausea, vomiting, an urge to defecate, chills, rapid breathing, confusion, and a tingling sensation in the hands or feet. To treat, pour water over the victim and fan him to speed up the coolant effect of evaporation, have the victim drink freshwater, and attempt to provide shade.

With heat stroke, the victim stops sweating, which results in red, flushed, hot, dry skin. The victim may first experience headaches, dizziness, nausea, fast pulse and respiration, and/or seizures and mental confusion. The victim can become unconscious and die if not treated quickly. To treat, create improvised shade for the victim. Have him drink water and elevate both legs if possible.

Burns

Apply a field dressing (or the cleanest material available) to the burn. Give sips of water to a casualty who is conscious and not nauseated. When treating burn casualties—

- DO NOT remove clothing stuck to the burns.
- DO NOT break any blisters.
- DO NOT apply grease or ointments to the burns.

For electrical burns, check for both an entry and exit burn from the passage of electricity through the body. An exit burn may appear on any area of the body, not necessarily opposite the entry burn.

For burns caused by wet or dry chemicals, flush the burns with large amounts of water and cover with a dry dressing.

For burns caused by white phosphorus, flush the area with water, then cover with a wet material, dressing, or mud to exclude air and keep the white phosphorus particles from burning.

For laser burns, apply a field dressing.

Common Medical Problems Associated with Sea Survival

Seasickness

Seasickness is the nausea and vomiting caused by the bobbing motion created by the wave action of a flotation device. Seasickness can result in—

- Dehydration and exhaustion.
- A loss of the will to survive.
- Others becoming seasick.
- Unclean conditions.

To treat seasickness—

- Wash both the Marine and the flotation device to remove the sight and odor of vomit.
- Keep the Marine from eating food until the nausea is gone.
- Have the Marine lie down and rest.
- Give the Marine seasickness pills if available. If the Marine is unable to take them orally, the pills should be inserted rectally for absorption by the body.

Saltwater Sores

Saltwater sores occur when skin that has abrasions or is cut is exposed to saltwater. The sores may form scabs and pus. Do not open or drain. Flush the sores with freshwater, if available, and allow to dry. Apply antiseptic, if available.

Blindness/Headache

Irritants or the effects of the sun's rays reflecting off the water can cause temporary blindness or headaches. If flames, smoke, or other irritants get in your eyes, flush the eyes immediately with saltwater, then with freshwater, if available. Apply an ointment, if available. Bandage both eyes for 18 to 24 hours or longer if the damage is severe. If glare from the sky and water causes your eyes to become bloodshot and inflamed, bandage the eyes lightly. Try to prevent this problem by wearing sunglasses or goggles with a sunglass insert.

Constipation

This condition is a common problem associated with dehydration. For constipation, do not take a laxative if it is available; this causes further dehydration. Drink freshwater, if available.

Sunburn and Dehydration

The sun's rays reflect at all angles off the waves of the water; therefore, sunburn and dehydration are serious problems in sea survival. Try to prevent sunburn by—

- Erecting an improvised canopy, with available floating materials, to provide shade.
- Wearing your soft cover or using a cloth, such as a handkerchief, to cover your head.
- Covering your skin with sunscreen or lip balm from your first aid kit. Your lips, nostrils, eyelids, the backs of your ears, and the skin under your chin sunburns easily. If enough sunscreen cream is available, all exposed skin should be covered.

Dehydration is caused by the loss of the body's vital fluids. Dehydration in saltwater may result from a combination of factors such as a lack of water, the effects of saltwater on skin tissue, sunburn or vomiting from seasickness, and other causes. Sleep and rest and reduced water and food intake are the best ways of enduring periods of exposure. The following measures will delay the effects of dehydration:

- DO NOT drink saltwater.
- DO NOT drink urine.
- DO NOT drink alcohol.
- DO NOT smoke.
- DO NOT EAT unless water is available.

Dangerous Marine Life

You may see many types of marine life around you, and some are more dangerous than others. Generally, sharks are your greatest danger, followed by barracudas. However, most marine life will not deliberately attack a human. The most common injuries from marine life are wounds from bites, stings, or punctures. With the exception of sharks and barracudas, most injuries are a result of either trying to catch game or from contact abrasion with marine life. To treat an injury that results from a dangerous marine animal bite—

- Pack the wound with gauze and then apply a pressure bandage, if available.
- Treat for shock.
- Obtain medical attention as soon as possible.

For a more detailed information on the treatment of casualties and dangerous marine life, see MCRP 3-02F and MCRP 3-02G.

Sharks and Barracudas

Only about 20 percent of all shark species are known to attack people. Sharks have an acute sense of smell, and the smell of blood in the water will draw them to their prey. They are also very sensitive to any abnormal vibrations in the water; therefore, the sound caused by a struggling swimmer or underwater explosions will attract them.

Barracudas are bold and inquisitive fish. They have been known to attack men who are wearing shiny objects. Barracudas may charge at lights or shiny objects at night.

A group of swimmers can maintain a 360 degree watch while in the water. Therefore, to protect yourself from sharks and barracudas, stay with other swimmers. The group members can either frighten away or fight off sharks or barracudas better than one person can. Keep all your clothing on, including footwear. Historically, sharks attack unclothed individuals in groups first, mainly in the feet. Clothing also protects you against abrasions from a shark's tough skin should the shark brush up against you. Avoid urinating heavily, let urine dissipate between discharges. If you must defecate, do so in small amounts and toss it as far away from the group as possible. Do the same if you must vomit.

If attacked, the use of firearms by swim sentries should be used with extreme caution because of the risk of injury to other swimmers. If unarmed or unable to make an improvised weapon, kick and strike the shark. Avoid using your bare hands to strike the shark, injury can result to your hands due to a shark's tough skin. Target areas on a shark are the gills, eyes, and underbelly. Blows to the snout are also not recommended because a shark will tilt its head up and thrust its jaws forward when biting.

Sea Snakes

Sea snakes are venomous and sometimes found in mid-ocean. They are unlikely to bite unless provoked. Your best protective measure is to avoid them.

Poisonous Fish

Many reef fish are toxic, and can kill you if eaten. Generally, poisons are present in all parts of the fish, but especially concentrated in the liver, intestines, and eggs.

Turtles and Moray Eels

Turtles and moray eels normally inflict minor bite wounds. Treat this type of injury by cleaning the wound.

Corals

Coral, dead or alive, can inflict painful cuts. Clean all coral cuts thoroughly. DO NOT use iodine to disinfect any coral cuts. Some coral polyps feed on iodine and may grow inside the flesh.

Jellyfish, Portuguese Man-of-War, Anemones, and Others

This group of marine animals inflicts injury by stinging their victims with their tentacles. Contact with their tentacles produces burning pain, a rash, and small hemorrhages on the skin. Shock, muscular cramping, nausea, vomiting, and respiratory distress may also occur. Gently remove the clinging tentacles with a towel and wash or treat the area. Use diluted ammonia or alcohol and talcum powder to treat the injury if available.

Spiny Fish, Urchins, Stingrays, and Cone Shells

These animals inject their venom by puncturing the skin with their spines. General signs and symptoms include swelling, nausea, vomiting, generalized cramps, diarrhea, muscular paralysis, and shock. Deaths are rare. Treatment consists of soaking wounds in hot water, if available, to deactivate heat-sensitive toxins.

Chapter 4
Negotiating Water Obstacles

Marines face water obstacles in saltwater, freshwater, and brackish water (where freshwater and saltwater meet). These water environments differ considerably and pose distinct problems for Marine tactical units and swimmers. Saltwater obstacles include tides, surfs, and currents. Brackish water obstacles include back bays. Freshwater obstacles include rivers and canals.

Tides

Tides are periodic changes in the surface levels of oceans, bays, gulfs, inlets, and rivers. The Moon's and Sun's gravitational pulls cause tides. Depending on the situation, tides can either help or hinder Marines in their endeavors to conduct amphibious or riverine operations. Tides can provide sufficient depths to allow for the passage of landing craft or boats over reefs, trees, and other underwater obstructions. On the other hand, tides can render a river fording site unusable to tactical vehicles. Direction, level of change, and amount of change determine tidal nomenclature.

Tides that show change in direction are flood tides and ebb tides. Rising tides are known as flood tides. Falling tides are known as ebb tides.

Tides that show extreme levels of change are high tides and low tides. High tide is the period when water is at its greatest depth. Low tide is the period when water is at its most shallow depth.

Tides that show the least and most amount of change are neap tides and spring tides, respectively. Neap tides have the least amount of change in water levels between high and low tide. Neap tides occur at the half moon, when the Sun and the Moon are aligned at a 90-degree angle with the Earth. In this position, the Sun's and Moon's gravitational pulls offset each other. Spring tides have the highest floods and lowest ebbs. Spring tides occur at or shortly after the new moon or full moon when the Sun, Moon, and Earth are approximately in line. In this position, the Sun's and Moon's gravitational pulls are combined.

Surf

Waves break upon entering shallow water and create surf. The offshore area where waves break is the surf zone, which can present many hazards. Breaking waves often trap air bubbles and create a foamy appearance. Bubbles lower the water's density and decrease buoyancy. Move through foamy surf as quickly as possible. The type of wave determines your survival technique.

Wave action moves you toward shore. Lie on your back or side with your head pointing in the direction of the beach and your feet pointing into the waves.

As one wave approaches the beach, another drains away; relax and do not swim against the draining water. When a new wave is within about 3 yards, start swimming toward shore. Continue to swim until the wave lifts you and moves you toward the beach. Once the wave loses forward momentum, relax and repeat the cycle. If nearing rocks, turn your body and approach feet first to reduce the chance of striking your head and arms.

Plunging Waves

A plunging wave is a breaker whose crest curls forward and falls ahead of its base. Because of its power and underwater turbulence, a plunging wave poses the greatest surf threat. If caught in a plunging wave, you can be pulled underwater and pitched about violently. This can cause you to panic, which can increase your chance of drowning. Perform the following steps to escape a plunging wave:

Tuck into a ball with your head against your knees and your forearms locked around your legs, just below the knees.

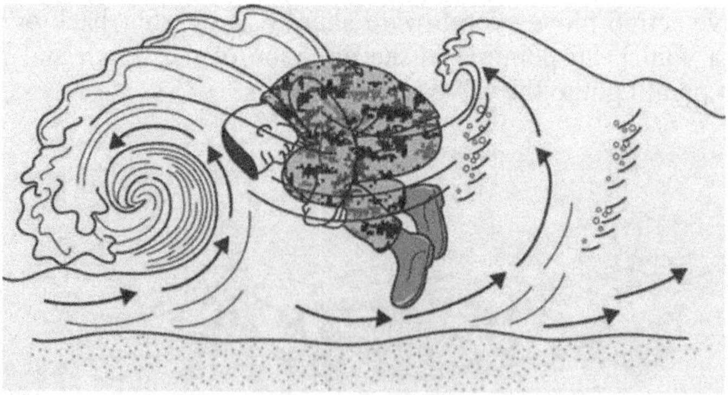

Relax in this position until the turbulence subsides and you float to the surface. This can take 30 seconds or more.

Swim toward shore.

NOTE: If threatened by another plunging wave, dive underwater into the wave.

Spilling Waves

A spilling wave does not break. Instead, its crest slides forward without curling. A spilling wave creates less turbulence and poses less of a threat than a plunging wave. If caught in a spilling wave, relax and let the wave carry you to shore.

Surging Waves

A surging wave occurs on a beach with a steep underwater gradient. It never really breaks, but the crest rises while the base slides up the beach with great force and speed. Once the wave reaches its highest point on the beach, it rushes back as quickly as it surged forward. If you are standing on the bottom when a surging wave advances or retreats, the wave can knock you off your feet and pull you into the surf zone. Do not try to stand or walk on the bottom. Swim toward the beach as soon as possible.

Currents

Offshore Currents

An offshore current occurs outside the surf zone. Typically, it occurs at bay entrances, in island channels, and between islands and the mainland. An offshore current flows parallel to or away from shore. If the offshore current is created by tides, its current strength and direction vary at different times of the day.

If caught in an offshore current, you may be carried in a direction you do not want to go. DO NOT try to swim directly to safety. If the current is moving directly away from the shore, relax and wait until the current dies out or turns toward land. Once the current subsides, use any survival stroke to swim toward shore. If the current is moving parallel to shore, use any stroke to move at an angle across the current and toward shore.

Rip Currents

A rip current occurs when waves pile water against the shore faster than the water can drain. The water flows rapidly along the beach until it is deflected seaward by a bottom obstruction. Then the rip current flows through the surf zone and into open water at a speed of up to 2 knots. This action can cut deep trenches in the sand. A rip current dies out once it reaches open water (usually within a few hundred yards of the shore).

A rip current can pull you out to the open sea. If caught in a rip current, DO NOT try to swim against the current. A rip current moves faster than most people swim, and it is impossible to swim to shore once caught in it. Relax and stay afloat until the current runs out. Once the current subsides, use a survival stroke to move parallel to the shore until you are out of the current, then begin swimming toward shore.

Littoral Currents

A littoral current occurs when a wave breaks against a beach at an angle. This current flows parallel to the shoreline and does not pose a great threat. If caught in a littoral current, use the combat travel stroke to swim across it at an oblique angle.

Back Bays

Once on the beach, you often face one or more rows of low hills called dunes. Behind the dunes, you may encounter a low-lying stretch of ground thickly covered with scrub trees and bushes. This area gives way to wetlands known as back bays. Back bays consist of muddy islands that are almost submerged during flood tides and separated by channels of brackish water of varying depths. Channel bottoms usually contain soft mud. Back bays pose major obstacles to vehicular traffic.

Infantry can cross back bays, but only with great effort. If crossing back bays by foot, consult detailed navigation charts and use the following guidelines to plan your route:

- Avoid water less than waist deep; walking in shallow water or soft mud is extremely tiring.
- Avoid back bay islands; these low-lying islands are usually too muddy to support foot traffic.
- Seek out deep water; floating with a pack is less tiring than walking through shallow water or soft mud.
- Seek out sand, shell, gravel, or stone bottoms; these firmer bottoms generally ease travel and help conserve energy.

Rivers and Canals

A river is a large, natural stream of water that empties into a larger body of water. The slope of the riverbed and the volume of water in the river determine its current.

Canals resemble small rivers or streams in their width and depth, but usually lack any significant current. Climbing out of these waterways can be difficult if the canal is flanked by steep banks.

Chapter 5
Fording Waterways

WARNING

Fords are dangerous. Cross as quickly as possible.

A ford is any site in a river, stream, or canal where the water is shallow enough for troops or vehicles to cross without using flotation devices. Canal bottoms are usually too soft to support fording vehicles, and wading infantry frequently stumble.

The tactical situation dictates the location of the fording site. Seek fords that are protected from enemy observation and that allow for adequate supporting fires. A night fording takes at least one and a half times as long as a daylight fording.

> *NOTE:* In this chapter, the term "river" refers to rivers, streams, and canals.

Selection of a Ford Site

The following table identifies desirable fording characteristics:

Characteristics	Comments
Concealment	The ford hides personnel and vehicle movement from enemy observation.
Accessibility	The ford should have low banks with gentle gradients. This allows a free flow of traffic at both the entrance and the exit.

Characteristics (Continued)	Comments (Continued)
Slow Current	The ford's current should not exceed 1.5 meters per second if possible.
Firm Footing	The ford's bottom, entry, and exit composition should be firm enough to support traffic. Do not drive a vehicle over any bottom composition that a 2-inch diameter stick can be pressed into more than 1 or 2 inches.
Gently Sloped Channels	The ford's entry and exit points should be gently sloped. If possible, locate a portion of the stream where the channel is not actively shifting.
Depth	The fording depth is less than or equal to the least capable vehicle.

Determine the Slope

Units move into and out of water faster and more quietly if entry and exit points are not steep or muddy. Slope is the amount of change in ground horizontal distance (run) and in vertical elevation (rise) from one point to another. Slope is usually expressed as a percentage. You can use a clinometer, map, or line of sight and pace to measure the percentage of a slope.

Clinometer

A clinometer measures percentage of a slope. It is a component of the M9 armored combat excavator, and organic to many Marine engineer units. However, engineers are also taught (and often use) field-expedient methods to determine slopes and gradients.

Map

A map measures the horizontal distance along a desired path. Determine the difference in elevation between the path's starting and ending points. From a map's scale, you can determine the distance between two points. From a map's elevation lines, you can determine the difference in height between the same two points. Both figures must be in the same unit of measure (e.g., feet, meter). Divide the elevation (rise) by the distance (run) and multiply by 100.

Rise = 165 - 120 = 45 meters

Run = 200 meters

% Slope = $\frac{45}{200}$ **x 100 - 22.5%**

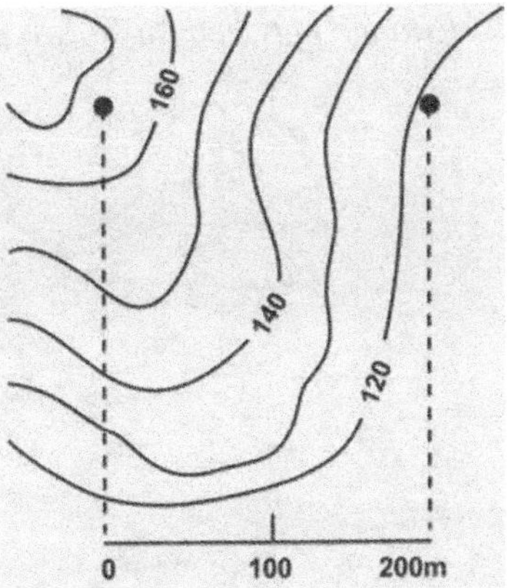

Line of Sight and Pace

To determine line of site and pace, stand at the bottom of the slope, keep your eyes level, pick a spot on the slope, then pace the distance. The number of paces multiplied by a standard measure of 0.75 meter determines the run. The eye-level height (usually 1.5 to 1.7 meters) determines the rise. Repeat this procedure until you have covered the entire distance you want to measure for each spot (vertical and horizontal). Add the vertical distances to provide total rise and the horizontal distances to provide total run.

Determine the Current Speed

Current speed increases as channels narrow. It may be necessary to locate a wider ford location to obtain a slower stream current. The following steps are used to calculate the speed of the current:

Determine points A and B along a channel. Then measure the distance between those two points.

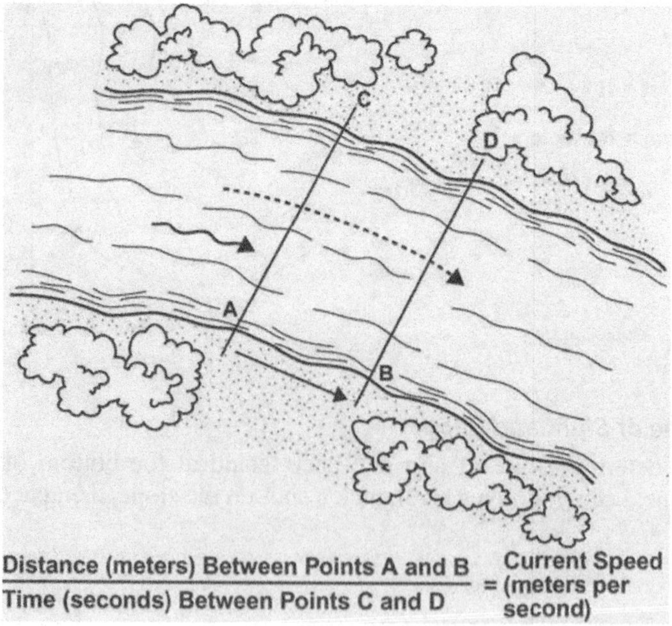

Sight directly across the water from points A and B to locate points C and D.

Throw a floating object (e.g., a stick) upstream from points A and C. Observe the object as it floats toward points B and D.

Subtract to find the time it takes for the object to float from start to finish.

Do not attempt to swim across currents that are moving faster than 1.5 meters per second. Equivalents of this speed include—

- Quick-time march rate of 120 counts per minute with one, 30-inch step at each count.
- 5 feet per second.
- 3.5 miles per hour.
- 5.5 kilometers per hour.

Measure River Width

A river's width can be estimated from the width of its symbol on a scaled topographic map. If this is not possible, use the following compass techniques:

Stand at the water line (A).

Shoot an azimuth to a point on the opposite bank (B).

Move upstream or downstream until you are at a point (C) where you can shoot an azimuth 45 degrees larger or smaller than the original azimuth.

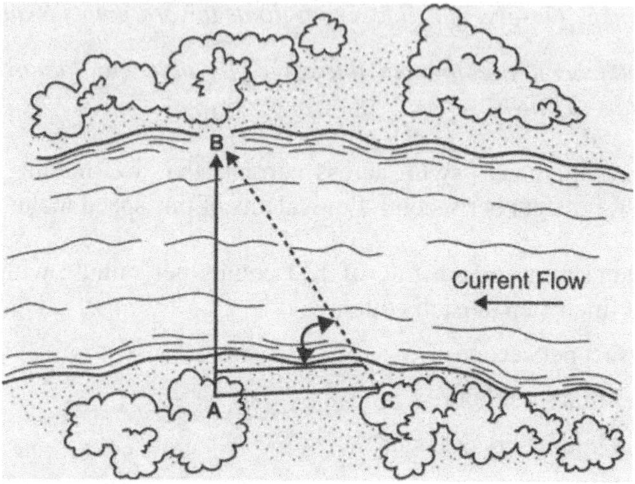

Measure the distance between points A and C. The distance calculated equals the river's width.

Calculate Downstream Drift

A river's current causes personnel and equipment to drift downstream. If personnel and equipment are aimed straight across the river, they will sideslip downstream as they move across the current to the other shore. Therefore, personnel and equipment crossings must compensate for the effects of a river's current and entries are usually made upstream of the desired exit point. Use the formula below to calculate downstream drift.

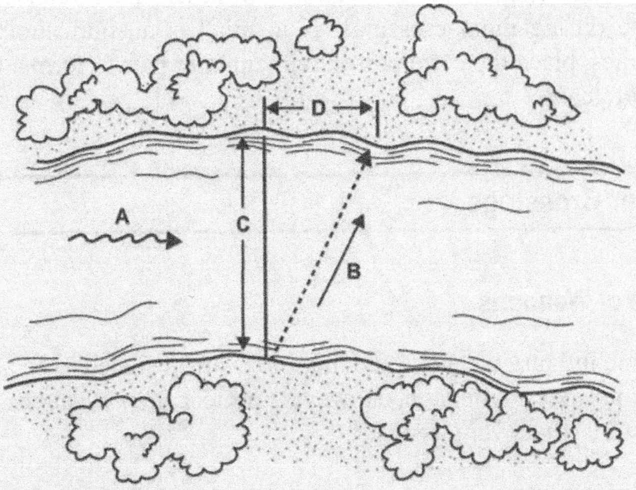

$$\frac{\text{Current Speed (A)}}{\text{Crossing Speed (B)}} \times \text{River Width (C)} = \text{Downstream Drift (D)}$$

> *Note:* The crossing speed for a swimmer across a river may vary but is generally limited to 1 meter per second. All measurements must be in the same unit of measure (e.g., meters, feet).

The Buddy System

Whenever a Marine unit must enter into or operate on the water, a "buddy system" is employed in which every Marine is paired with a swimming partner. The buddy system matches an experienced swimmer with a weak swimmer. The experienced swimmer assists and encourages the weaker swimmer and bolsters confidence during night crossings. If a unit has an odd number of Marines, place the extra person with another pair to form a three-person team.

Water Crossings

Care of Weapons

Marine infantry weapons and munitions are designed to be able to operate after immersion. However, protect your weapons from moisture whenever possible.

A gas-operated weapon can malfunction if water travels down the barrel and enters the gas tube. To protect the gas tube—

Close the weapon's bolt before entering the water.

Seal the muzzle with a condom, balloon, plastic spoon wrapper, or other form of waterproof material.

Tie or melt the protective cover to create a watertight seal.

When the muzzle's protective cover is no longer needed, remove it. Open the bolt and inspect the barrel. If the tactical situation permits, swab excess moisture from the barrel. Test fire automatic weapons, if possible. Field strip and clean weapons as soon as possible. If time does not allow for a complete inspection, rinse inaccessible areas with small amounts of diesel fuel, then dry.

ISOMAT Raft

Construction of an ISOMAT raft is time-consuming. This type of raft should not be employed as part of an attack, but used for logistical purposes (e.g., evacuating stretcher cases, transporting supplies). Use the following steps to build an ISOMAT raft:

Wrap ISOMAT sleeping pads around sturdy sticks.

Use parachute cord and square knots to tie the pads securely in place and to lash stick ends together in a rectangle.

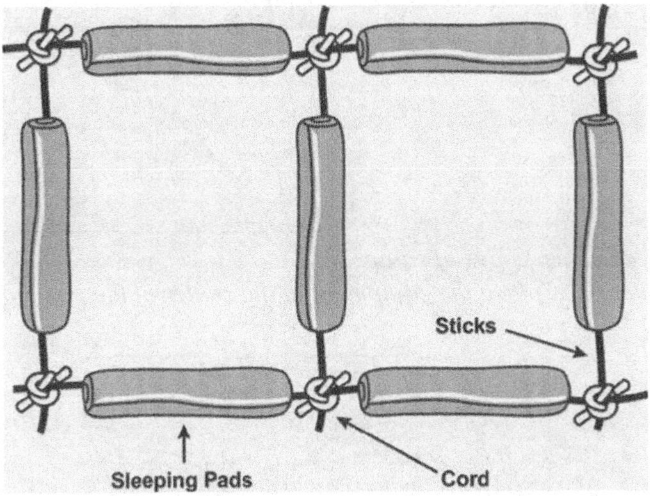

Note: The ISOMAT raft pictured can support several hundred pounds. However, the cargo will get wet if not properly waterproofed.

Poncho Raft

A poncho raft can support two Marines and their equipment and is well suited for long crossings. Use the following steps to build a poncho raft:

Inspect two ponchos and ensure they are serviceable.

Lay one poncho flat on the ground, with the hood-side up.

Cinch the hood tightly to form a gooseneck or tie in a knot.

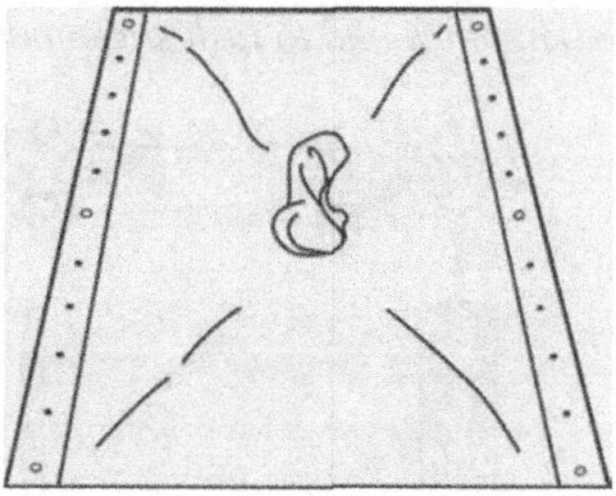

Pad sharp edges of equipment and place the equipment in the center of the poncho.

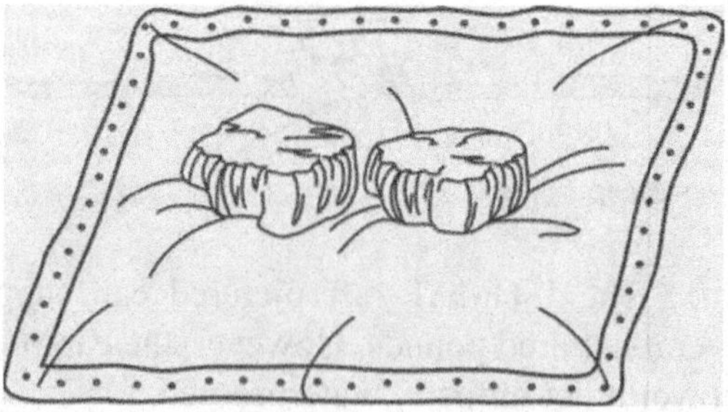

Place the second poncho over the equipment, rubber side up, and hood facing down.

Snap the edges of the two ponchos together.

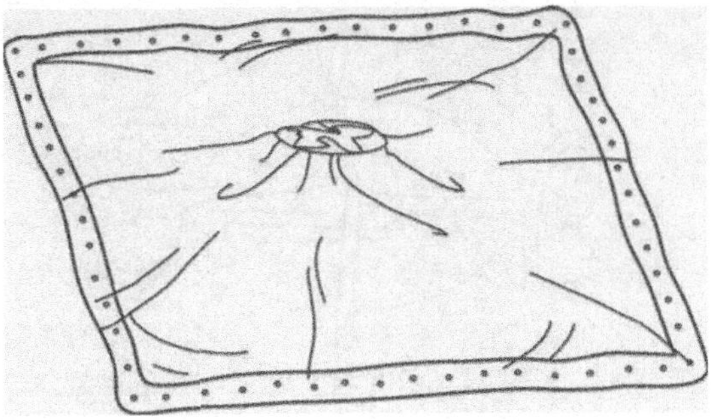

Roll the edges toward the equipment.

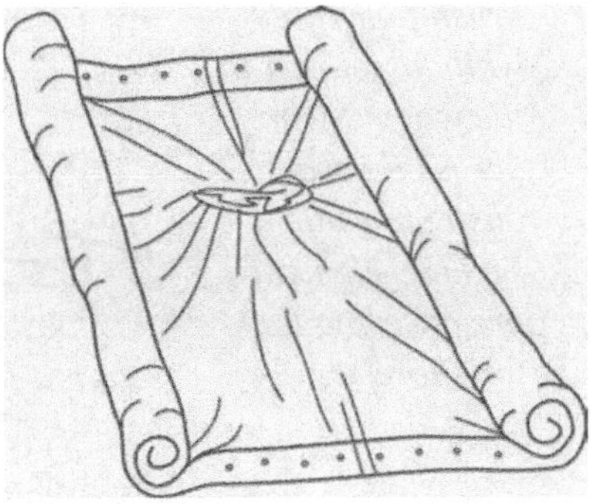

Roll the edges into pigtails and tie them off.

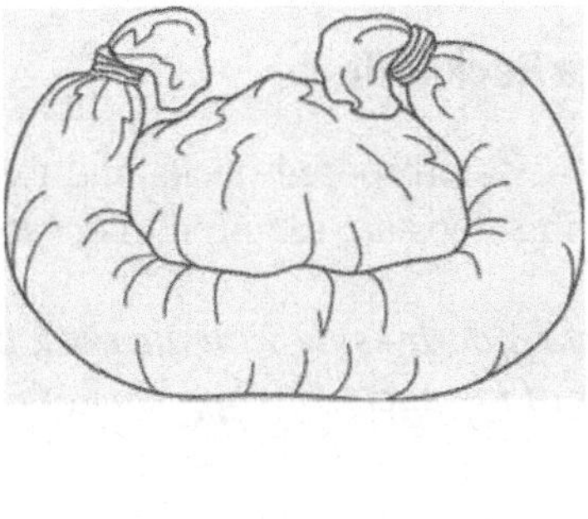

Pull the pigtails together over the top and lash them securely.

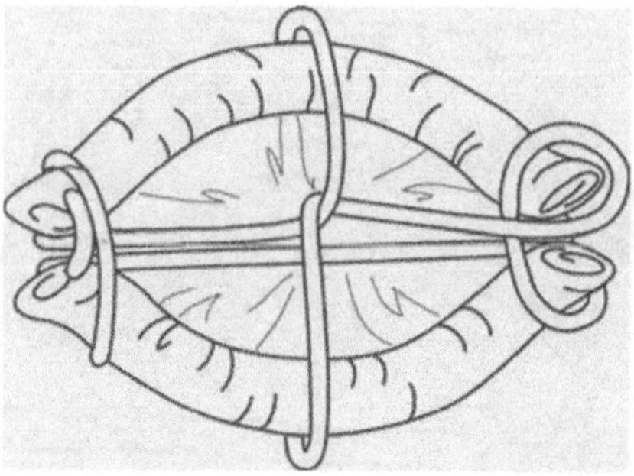

Protect the raft from brush punctures while placing it in the water. Swim across the water obstacle while security elements are covering the far shore.

Construction of a Pack Raft

You will need two waterproofed packs and two M16A2 service rifles. The following steps are required to construct a pack raft:

Place two packs side-by-side with the pack frames on the deck. The tops of the packs are opposite of each other.

Loosen the main compartment straps on both packs.

Insert one rifle on each end between the straps and the packs. The muzzles are opposite of each other. The rifles serve as one means to secure the packs together. Place the front sight post under the top flap.

Tighten the straps so that the rifles and packs are secure.

Take the excess strap on the inner side of each pack and secure it to the opposite pack to better secure the two packs together.

Take the excess straps on the outer sides of the packs and use those straps as safety lashing for the rifles.

Tuck the excess straps and check to make sure the rifles and packs are secure.

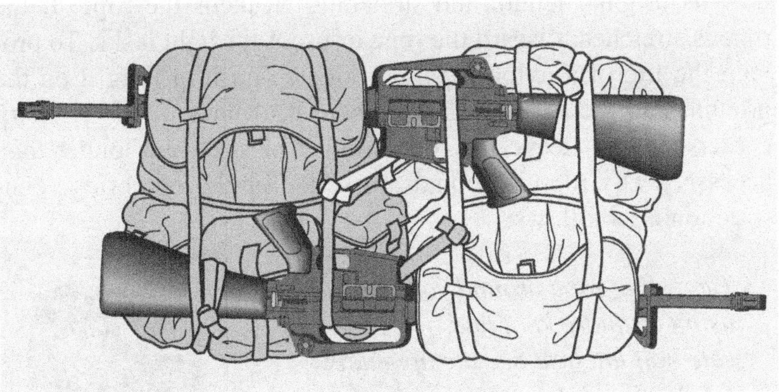

Single-Rope Bridge

A single-rope bridge offers a temporary and quick way to cross small rivers. It also provides extra security while crossing swift waters. At night, it prevents straggling, and guides units precisely from one side of the river to the other side. If crossing a river at night, plan for at least one single-rope bridge.

If your unit is crossing a river with swift currents or water depths above 4 feet, the unit is carrying sufficient rope to span the crossing site, and the tactical situation permits, secure the rope on near and far banks to provide a hand-hold for crossing Marines. This reduces the time required for the entire unit to cross and provides a degree of comfort/confidence for poor swimmers. Use a squad-sized bridge team to construct a single-rope bridge. Station several strong swimmers at the water's edge to help anyone who has trouble crossing.

Nylon rope is normally coiled in 120 foot lengths. It is 0.6 inches in diameter and has a breaking strength of about 3,840 pounds. Over time, a nylon rope can stretch to as much as one-third more than its original length and stretching weakens the rope. If the rope is stretched, discard the rope or use it for light tasks. To prolong the life of a nylon rope, do not step on it or drag it on the ground. Pad the rope in places where it contacts rocks or sharp corners. Do not leave the rope knotted or stretched longer than necessary. Dry rope as soon as possible. Single-rope bridge construction is as follows:

Tie a sling rope around your waist using a square knot and two, separate half hitches. See the appendix for detailed information on knots.

Attach a locking steel carabiner to the sling rope.

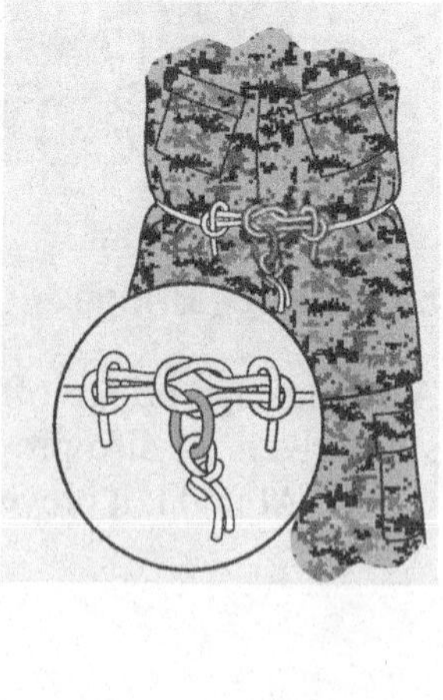

Tie a bowline knot in the running end of the bridge rope and attach it to the carabiner.

Temporarily secure the other end of the rope to a tree on the near shore.

Enter and cross the water.

NOTE: Carry only your weapon and ammunition.

Exit the water on the opposite shore.

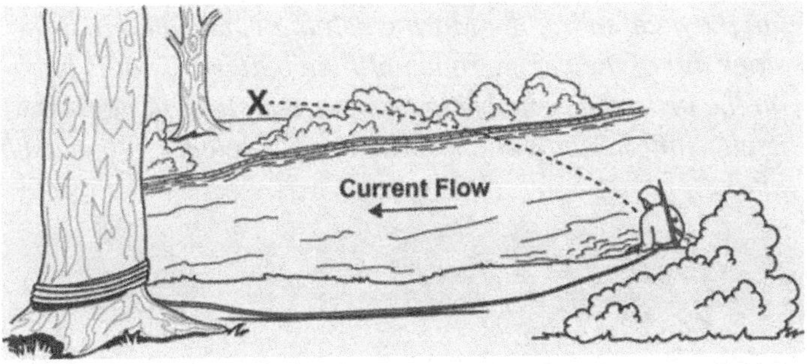

Prepare your weapon for use. Unhook the bridge rope from the carabiner at your waist, and tie the bridge rope to a sturdy tree using a round turn and two half hitches.

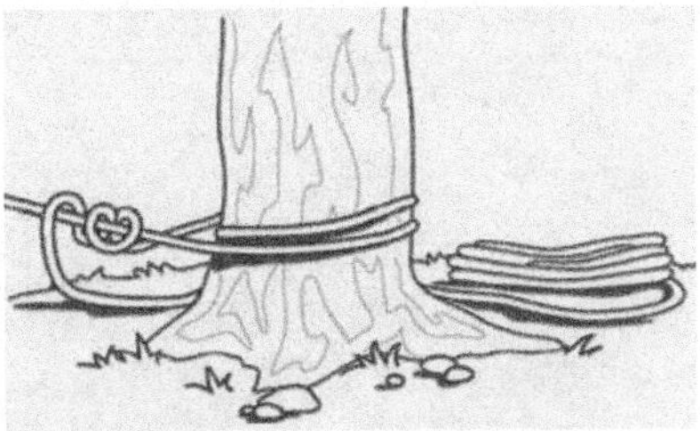

Conduct a box reconnaissance of the opposite shore.

On the near shore, have another Marine prepare to tighten the rope. That Marine should place a transport tightening system in the bridge rope by tying a double butterfly knot and placing two carabiners in the butterfly. See the appendix for detailed information on knots.

The Marine should pass the running end of the bridge rope around the downstream side of the near shore anchor point and through the two carabiners.

Pull the butterfly knot approximately one-third of the distance across the river.

Secure the bridge rope to an anchor point using a round turn and two half-hitches.

On the near shore, the Marine helping you should pull the slack out of the bridge rope until the butterfly knot is back on the near side. The bridge rope is then tied off against itself using two half hitches with a quick release in the last half hitch.

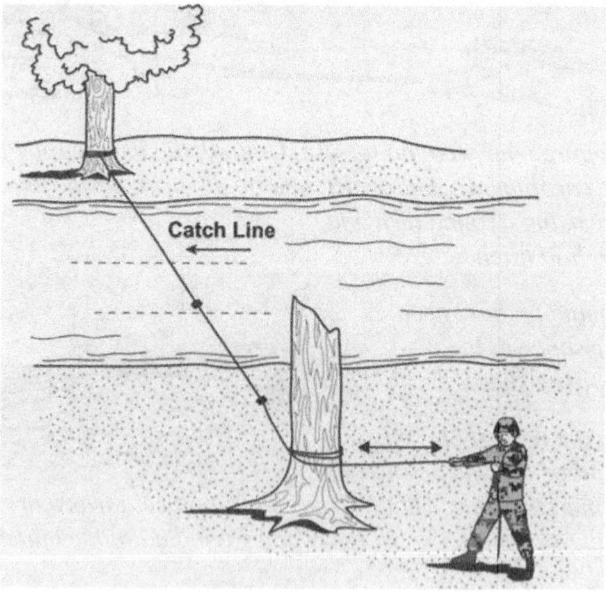

Note: The single-rope bridge must be as tight as possible so it will not sag when used.

If you lose your footing and fall into the water, swim with the current to the closest shore. Swimming against the current is dangerous and quickly causes fatigue.

High and Dry Crossings

If the single-rope bridge is high enough, suspend yourself below the single-rope bridge and above the water. Use the following steps to suspend yourself from a single-rope bridge and then pull yourself across the water:

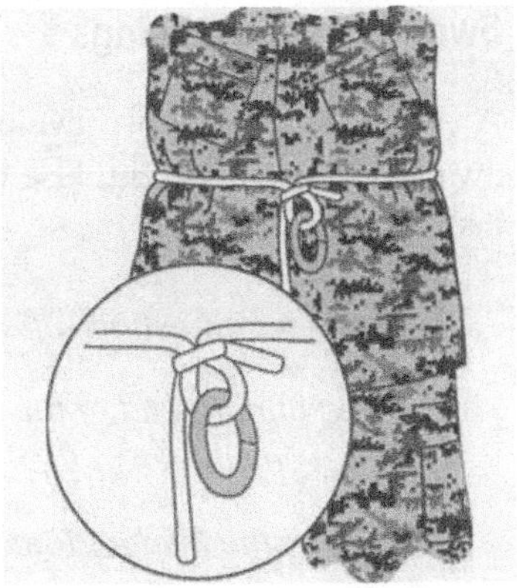

Tie a sling rope around your waist using a bowline. Ensure that the knot is tight.

Attach a carabiner through the bowline's loop. The carabiner's gate faces up.

Secure your helmet chin strap.

Face the single-rope bridge with your left shoulder toward the far shore.

Grasp the bridge rope in both hands.

Swing your body beneath the single-rope bridge with your head toward the far shore. Cross your ankles above the bridge rope.

Arch back until the carabiner contacts the bridge rope. Connect the carabiner to the bridge rope. Allow the carabiner to bear your body's weight.

Pull yourself across the single-rope bridge, hand over hand, to the far shore.

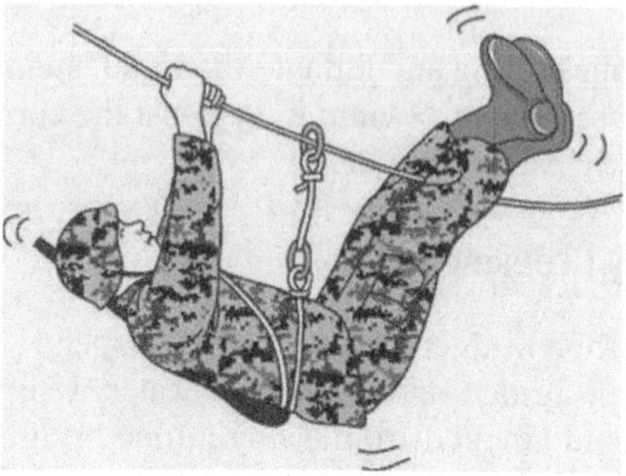

Swift Current Crossings

A single-rope bridge prevents being knocked down and swept away by a swift current. Use the following steps to move through a swift current:

Tie one end of a sling rope around your waist using a bowline.

Tie the running end of the sling rope in another bowline, and attach a carabiner to the bowline's loop.

Step up to the bridge. Face upstream.

Hook the carabiner to the single-rope bridge.

Walk sideways into the river while grasping the bridge rope in both hands.

Use the single-rope bridge for balance and remain standing, if possible.

Continue to move sideways through the river to the far shore.

Slow Current Crossings

If you face little or no current, it is not necessary to hook up to a bridge rope with a carabiner. Lie on your back in the water beneath the single-rope bridge. Support your body weight with your waterproof pack. Use the bridge rope and pull yourself hand over hand across the river.

Removal

If you are the last Marine waiting to cross, pull on the standing end of the rope to release the knot, then tie the rope around your waist using a bowline. The Marines on the far shore will pull you through the water.

Appendix
Knot Tying

Square Knot

The square knot is used to secure two ropes of equal diameter together so they form one continuous rope that will not slip.

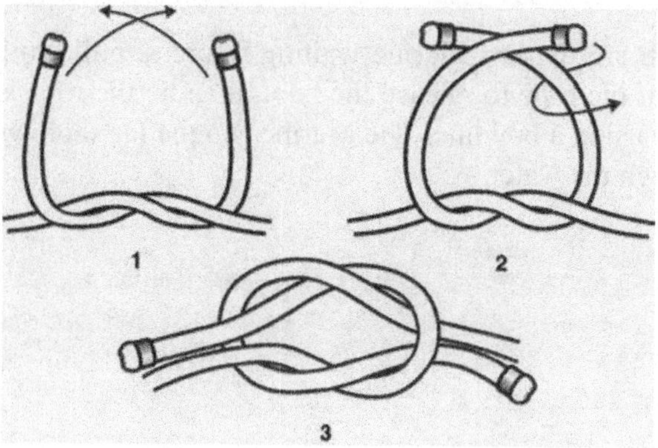

Bowline

The bowline forms a loop that will not tighten or slip under strain. It is easily untied.

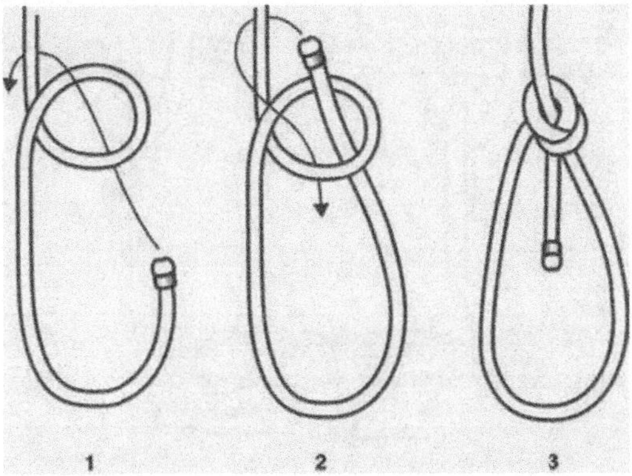

Hitches

Hitches are used to dress (prepare) knots and secure loose ends.

Half Hitch

The half hitch is used to tie a rope to a tree or to a larger rope. It will hold against a steady pull, but is not a secure hitch. It is frequently used to secure the free end of a rope.

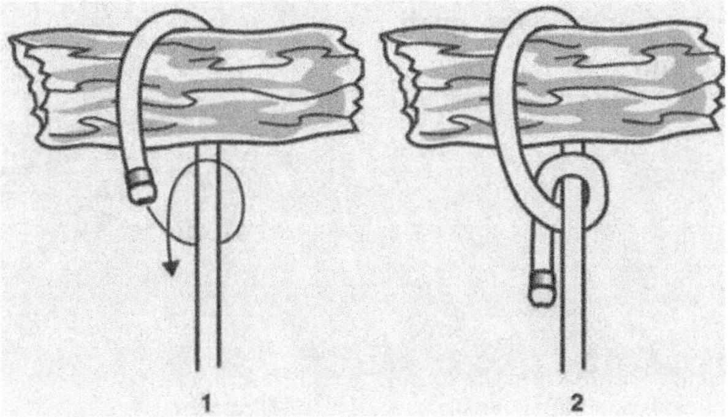

Two Half Hitches

Two half hitches can be used to secure the running end of a rope.

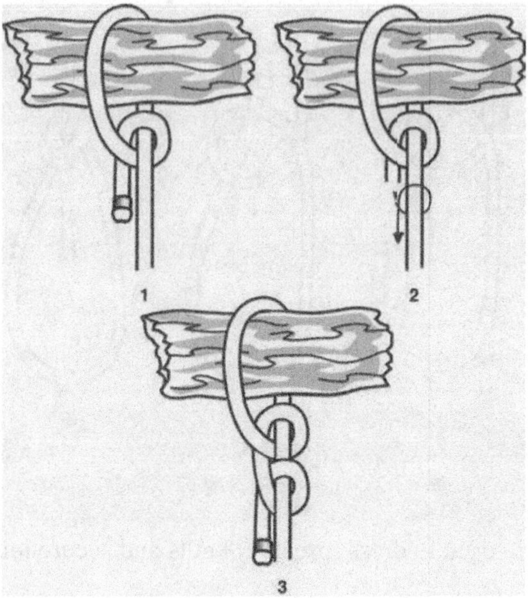

Round Turn and Two Half Hitches

A round turn and two half hitches can be used to fasten a rope to a tree. This hitch does not jam.

Butterfly Knot

The butterfly knot is used for anchor lines. Tied properly, it will not tighten on itself to the point that it cannot be easily untied. The butterfly knot can be used to pull a rope bridge taut. This knot can be used to tighten a fixed rope when mechanical means are not available. It will not jam if a stick is placed between the two upper loops.

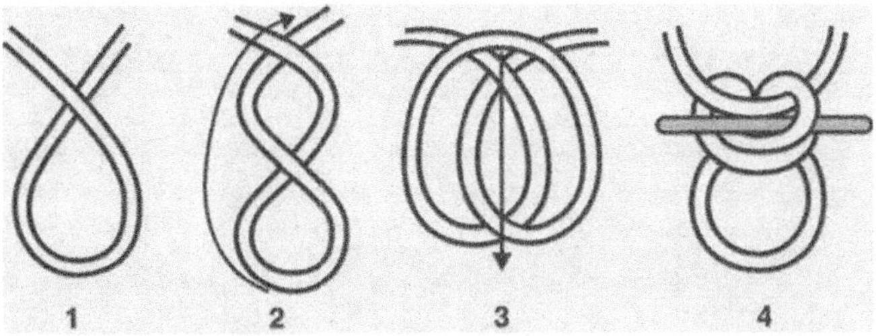

CONTENTS

INTRODUCTION

0001. PURPOSE. To promulgate policy and implementing instructions, concerning physical fitness and body composition in the Marine Corps.

0002. STATUS

1. The policies and procedures in this Manual apply to HQMC staff agencies and COs who are authorized to issue directives concerning this program.

2. The CMC (C 472TP) must authorize any deviation from the instructions in this Manual.

0003. RESPONSIBILITY. The CMC (C 472TP) is responsible for the accuracy, currency, modification, and distribution of this Manual. HQMC staffs and COs are responsible for the timely entry of changes and the physical maintenance of copies of this Manual.

0004. ALLOWANCES

1. Submit requests for changes in allowances of this Manual to the CMC (C 472TP). Submit requests per reference (b).

2. Request missing pages by requisitioning the basic Manual and pertinent changes per reference (b).

0005. ORGANIZATION

1. This Manual is organized into chapters identified by an Arabic numeral as listed in the overall contents.

2. Paragraph numbering is based on four digits. The first digit indicates the chapter; the next digit, the section; the final two digits the general major paragraph number; and the combinations which follow the decimal point, the subparagraph number; e.g., 3103.3a(2) refers to chapter 3, section 1, general major paragraph number 03, subparagraph 3a(2).

3. Pages are numbered in separate series by chapter number, with the chapter number preceding each page number; e.g., the fourth page of chapter 2 is shown as 2-4.

4. A glossary of terms is provided in appendix A to assist users with definitions of the numerous terms and acronyms throughout this Manual.

0006. CHANGES. Printed changes are posted per reference (c). Interfile changes in the basic Manual in consecutive order and complete the Record of Changes page.

CHAPTER 1

PHYSICAL CONDITIONING PROGRAM (PCP)

CHAPTER 1

PHYSICAL CONDITIONING PROGRAM (PCP)

1100. SCOPE. Every Marine must be physically fit, regardless of age, grade, or duty assignment. Fitness is essential to the day-to-day effectiveness and combat readiness of the Marine Corps. Furthermore, physical fitness is an indispensable aspect of leadership. The habits of self-discipline required to gain and maintain a high level of physical fitness are inherent to the Marine Corps way of life and must be a part of the character of every Marine. Marines who are not physically fit can be a detriment to the readiness and combat efficiency of their unit. Accordingly, every Marine will engage in an effective PCP on a continuing and progressive basis.

1101. RESPONSIBILITY. In order to maintain continuity of standards, the procedures in this Manual will be used for developing and administering the COs conditioning, testing, and remedial programs, and to clarify Body Composition Program (BCP) policies (formerly known as the Weight Control Program). It is the responsibility of COs to provide a fair evaluation of Marines' physical fitness and body composition through the procedures outlined in this Manual. When necessary, an Appropriately Credentialed Health Care Provider (ACHCP), which will be referred to throughout this Manual as a Medical Officer, or "MO", will be required to conduct a thorough physical evaluation; an ACHCP is defined in the Glossary, appendix A of this Manual. This evaluation will include at a minimum, a Body Composition Evaluation and a determination of whether or not the weight and body fat gain is due to an underlying cause or associated disease.

1102. OBJECTIVES

1. Overall Program Focus. The Unit's PCP should focus on combat conditioning, health, fitness, and unit cohesion rather than preparation exclusively for the semi-annual Physical Fitness Test (PFT). The Program should reflect the following objectives:

a. To contribute to the health and well-being of every Marine through regular exercise, fitness testing, and health education.

b. To develop Marines who are physically capable of performing their duties in garrison and in combat.

c. To develop in Marines a reserve level of physical fitness and endurance that will enhance their chances of success in a combat situation.

d. To provide a medium for developing the individual Marine's self-confidence and desire to excel, thereby enhancing the unit's overall discipline, morale, esprit de corps, and combat readiness.

2. CO's Focus. COs should strive to achieve the primary objective of combat readiness by emphasizing combat conditioning at the unit level. The CO should ensure this objective is reflected in the training plans and schedules, affording the individual Marine an opportunity to achieve a competent level of combat fitness.

3. Command Physical Training (PT) Representative's Focus. The CO will assign in writing a Marine SNCO (normally from the Operations and Training Section), as the Command PT Representative. This Marine will be the commander's senior representative in the administration of Marine Corps policy related to physical conditioning, testing and remedial programs. Of particular emphasis, this Marine will be responsible for:

a. Conduct of the PFT. Responsible to the CO for selection and training of PFT monitors and for all matters related to the PFT (safety, course selection, recording, etc.).

b. Body Composition Evaluations. To ensure accuracy in body composition evaluations, the **only** personnel authorized to conduct these measurements are the Command PT Representative and his/her designate.

1103. PHYSICAL CONDITIONING. The definition of physical fitness is subjective to those defining it. The Marine Corps considers physical fitness to be the ability of a Marine to meet the physical demands of any combat or duty situation without undue fatigue. To achieve this state of physical fitness, physical conditioning should incorporate the components and principles listed below.

1. Components of Physical Conditioning. To capitalize on those components that can benefit conditioning efforts, the following categories of exercises should be included in both individual and unit PCPs:

a. Strength. Muscular strength refers to the ability of the muscular system to move the body through resistance. Many associate strength training with progressive resistive exercises using weights and machines. However, the ability of a Marine to effectively handle their own body weight should be a prerequisite before integrating strength training with machines into their program. This can be accomplished through the Daily 16 Program described in paragraph 1103.3a, and applying the principles listed in paragraph 1103.2. Strength training can be broadly separated into two categories, general and specific.

(1) General Strength Training. This type of training strengthens the muscular system by focusing on a full body workout for strength and size. In this type of training, the major muscle groups are exercised without a specific task or functional goal in mind. This type of strength training contributes to overall health.

(2) Specific Strength Training. This type of strength training is task specific. For example, Marines desiring to climb a rope better would do rope climb training wearing body armor, and focus their strength training on muscles involved in rope climbing. A company of Marines expecting to operate in hilly terrain would focus their strength training primarily on lower body strength.

b. Endurance. Two types of endurance conditioning are needed for a Marine to meet the physical demands of combat, aerobic and anaerobic.

(1) Aerobic Endurance. Aerobic activity, meaning "in the presence of oxygen," is categorized by physical demands that are sub-maximal (not an "all out" effort) and involve activity that is continuous in nature (lasting more than 3-5 minutes). Two examples are road marching and long distance running.

(2) Anaerobic Endurance. Anaerobic activity, meaning "without oxygen," is categorized by physical demands that are high intensity and of shorter (less than 2-3 minutes) duration. Examples are rope climbing, most forms of weight lifting, and running short, quick distances.

c. Mobility. Mobility conditioning is geared towards improving quality of movement. Quality of movement depends on the following:

(1) Posture
(2) Balance and stability
(3) Agility
(4) Coordination
(5) Power
(6) Speed
(7) Flexibility

2. Principles of Physical Conditioning. There are several different principles to consider when developing an effective PCP:

a. Progression. Conditioning programs must incorporate a systematic means to increase training load.

b. Regularity. To realize a conditioning effect, training programs must conduct physical conditioning sessions at least 3-5 times per week.

c. Overload. Only when the various systems of the body are overloaded will they become able to handle greater load.

d. Variety. Varying a program from time to time maintains interest and prevents staleness.

e. Recovery. Essential for allowing the systems overloaded during conditioning to adapt and become stronger.

f. Balance. Balanced conditioning programs ensure all the components of physical fitness conditioning (strength, endurance, and mobility) are properly addressed.

g. Specificity. Conditioning that is specific in nature yields specific gains. For example, stationary bike riding is of little value in improving running.

3. Specific PCPs. Units that strive to augment their PCP should use innovative combinations of the types of exercise defined below:

a. Daily 16 Program. The Daily 16 Program is a comprehensive series of warm-up, conditioning, and cool-down exercises replacing the former Daily 7 Program. This all-encompassing program can be incorporated into any unit aerobic or anaerobic conditioning session, or can be used as a conditioning session in of itself. Appendix B contains the Daily 16 Program in detail.

(1) Warmup Exercises. The Daily 16 warm-up exercises facilitate gradual distribution of blood flow to the muscles, preparing both the cardiovascular and musculoskeletal systems for the exercise session, by effectively targeting both the upper and lower body. The increased blood flow to the muscles produces a warming effect, increasing the elasticity of the muscles and connective tissue, which is believed to reduce injury.

(2) Conditioning Exercises. The Daily 16 conditioning exercises provide a total body workout through the proper execution of traditional calisthenics. Conditioning exercises can be used as a workout session in and of itself, or to augment the main fitness event (e.g., squad ability run or obstacle course).

(3) Cool-down Exercises. The Daily 16 cool-down exercises (which are basically the same as the warm-up exercises) allow the body to gradually return to the pre-exercise state.

b. Occupational Conditioning. This conditioning is comprised of general physical conditioning exercises that will develop and maintain strength, endurance, and the physical skills necessary to sustain a Marine during combat. Good examples of occupational conditioning that prepares Marines to successfully handle the demands of their particular MOS/billet are Physical Readiness Training (PRT), progressive load-bearing marches, martial arts training, dry net training, Military Operation in Urban Terrain (MOUT) training, combat water survival training, obstacle course, and confidence course.

c. Competitive Conditioning. Competitive conditioning activities consist of teams or individuals competing against an opponent to win. This includes a combination of sports and military skills designed to foster the unit's combat readiness, competitive spirit, and esprit de corps. Some good examples are speed-marches reaction courses, orienteering, water-can re-supply relays, and stretcher or fireman carry relays.

d. Remedial Physical Conditioning Program (RPCP). All Marines who fail the PFT, demonstrate poor physical performance in MOS-related skills, or who are assigned to the BCP will be assigned to a RPCP. Remedial physical conditioning is a process by which Marines are conditioned in a progressive manner to meet prescribed standards. The goal is to provide challenging conditioning sessions, using the spectrum of aerobic conditioning, resistance conditioning, and other related exercises that will afford the Marines an opportunity to rebuild themselves after a weight and body fat increase, injury/illness, pregnancy, or a period that lacked a structured fitness program. The philosophy of remedial conditioning is not punitive in nature, but is intended to encourage improvement in overall fitness or to supplement a BCP. Comprehensive guidance for the development of a command directed RPCP is contained in paragraph 2400 and appendix C of this Manual.

e. Alternate Aerobic Conditioning. In cases of lower limb injury or related medical conditions that prevent running or hiking, Marines may opt to perform other low impact activities to supplement the requirement for cardiovascular conditioning. Swimming, deep water running, cycling, cross-country skiing, stair stepping, and rowing are excellent examples of low impact, endurance exercises that provide good augmentation or substitution for running or hiking regimens. For example, the advantages of swimming include the use of all major muscle groups during exercise, reducing lower body stress due to partial body weight being supported by the water, and the absence of impact on bones and joints. Additionally, the body's position in the water increases the efficiency of the circulation back to the heart during exercise. Although these types of aerobic conditioning alternatives may not improve overall running ability (e.g., speed and endurance), it can enhance a current running or hiking regimen, while minimizing related injuries.

1104. MARINE CORPS PCP REQUIREMENTS

1. Individual Requirements. The PCP requires every Marine to:

a. Participate in a minimum of 3 hours of physical fitness conditioning per week; 5 hours is encouraged if the unit's operation tempo can accommodate such a program.

b. Be tested semi-annually - PFT. (Reserves tested annually.)

c. Obtain a minimum 3rd Class score on all semi-annual PFTs.

1105. CO's ACTIONS. The PCP must provide opportunities for development through supervised individual or group effort, and sustained by common sense leadership at all unit levels. In order to develop and maintain the desired level of fitness, COs will:

1. Schedule a minimum of 3 hours per week of mandatory individual or group physical conditioning sessions. COs are encouraged to use normal working hours to satisfy this training requirement. It is recommended that each conditioning session be a minimum of 60 minutes in duration.

2. Establish, implement, and maintain the PCP requirements as stated in this Manual.

3. Ensure all Marines participate in exercises or activities commensurate with their medical qualifications and limitations.

4. Ensure proper administration and tracking for all Marines in a light and limited duty status (e.g., Morning Reports).

5. Ensure the unit's PCP is not constructed and implemented solely to prepare for a PFT, but that it emphasizes physical conditioning for combat. This distinction should be reflected in the unit training schedules.

6. Conduct a semi-annual PFT for all Marines per the instructions contained in chapter 2 of this Manual.

7. Place and track all Marines who fail the PFT on a RPCP per the guidance set forth in paragraph 2400 and appendix C.

8. Immediately process Marines for a BCP assignment when identified as not in adherence with weight and body fat standards and do not meet Physical Performance Evaluation criteria.

9. Place and track all Marines assigned to a BCP on a RPCP for the duration of their 6-month BCP assignment.

10. Employ Semper Fit resources to assist in the execution of the PCP, RPCP, and BCP. Semper Fit Fitness Centers staffed by qualified professionals, and the Semper Fit website (www.usmc-mccs.org) provide expert guidance in the areas listed below. The fitness centers, available at most major Marine Corps installations, can provide the following services:

a. Fitness assessments and exercise prescriptions*;

b. Exercise orientation;

c. Group exercise opportunities;

d. Instructional skill development; and

e. Health promotion programs.

*Note: Personal fitness assessments conducted by Semper Fit personnel (e.g., body fat testing), while valuable, will not be considered official for BCP testing.

11. Ensure timely and accurate reporting of PFT results and BCP assignments on fitness reports per reference (d), and that correct Unit Diary and SRB entries are made, as appropriate, per references (e) and (f).

12. Ensure the Command retains all completed PFT/BCP files for one year.

CHAPTER 2

PHYSICAL FITNESS TEST (PFT)

SECTION 4: REMEDIAL PROCEDURES

TABLES

CHAPTER 2

PHYSICAL FITNESS TEST (PFT)

2000. GENERAL INFORMATION

1. Purpose. While units are responsible for testing their Marines for combat fitness and readiness as it pertains to their unit's mission, the PFT provides an instrument that measures the collective level of physical fitness Marine Corps wide. It is a measurement of general fitness vice combat readiness and unit/MOS capability. The PFT consists of three events: male Marines will perform dead-hang pull-ups, abdominal crunches, and a 3.0 mile run; and female Marines will complete the flexed-arm hang, abdominal crunches, and a 3.0 mile run. These events are designed to test the strength and stamina of the upper body, midsection, and lower body, as well as the efficiency of the cardiovascular system.

2. Personnel to be Tested. Units will administer the PFT semi-annually (calendar year) for score to all active duty Marines, regardless of age, gender, grade, or duty assignment.

a. PFT Not Required. The condition when a Marine will not be required to conduct a semi-annual PFT:

(1) Personnel on duty in a combat zone; the semi-annual PFT may be waived until their return from the combat zone.

(2) Pregnant Marines medically waived from taking a PFT.

(3) Marines in a light duty or limited duty status, restricting them from participating in the PFT or Partial PFT (PPFT). (Appropriate documentation by a Medical Officer (MO), that assigns light duty or directs a Medical Evaluation Board (MEB) action, is mandatory.)

(4) Marines awaiting separation based on a Physical Evaluation Board (PEB). Reference paragraph 2302 for further details.

b. PFT Requirement for End of Active Service (EAS)/Retirement

(1) Marines are required to take the PFT during the semi-annual period preceding their EAS or retirement date, unless directed by their CO. This date is not the date that terminal leave begins.

For example, a Marine whose EAS or retirement date is September, but is commencing terminal leave in June, will be required to take the semi-annual PFT for the January-June period. This Marine is not required to take the semi-annual PFT for the July-December period.

(2) Retirement/EAS Final Physical Examination. Completion of the required final physical examination, regardless of when it is completed, does not exempt a Marine from taking the PFT. A Marine, who elects to complete their final physical examination 7-12 months prior to EAS or retirement, will still be required to take the semi-annual PFT for that period.

For example, a Marine who completes the final physical in March, but does not EAS or retire until October, will still be required to take the semi-annual PFT for the January-June period.

c. Selected Marine Corps Reserve (SMCR). Reserve Marines are required to conduct an annual PFT for score. Reserve Marines on Active Duty Special Work (ADSW) orders for 30 days or more, can conduct their annual PFT with that unit. All scores are good for 2 years (for cutting score purposes only), if training does not permit testing.

CHAPTER 2

PHYSICAL FITNESS TEST (PFT)

SECTION 1: TESTING PROCEDURES

2100. CONDUCT OF THE PFT

1. Uniform. The authorized uniform for the semi-annual PFT is the USMC established green-on-green T-shirt and shorts, socks, and running shoes. The green-on-green sweat suit may be used in winter climates. Gloves may be worn. Substitutions of the aforementioned uniform are at the discretion of the CO.

2. Sequence of Events. The sequence of PFT events will be left to the discretion of the CO. All PFT events will be conducted in a single session, not to exceed 2 hours in duration. Movement of Marines from one event to the next should allow adequate time to recover, stretch, and drink water.

3. Restriction of Events. The command will not impose restrictions on the performance of any event other than those specifically stated in this Manual. For example, a command will not mandate that Marines must use the overhand grip when executing pull-ups or flexed-arm hang.

4. Monitoring. The Command Physical Training (PT) Representative will monitor the PFT and maintain the PFT and Body Composition Worksheets. Marines will be kept advised of their progress as they are performing each event. The Command PT Representative or designate will call out each repetition for the pull-ups, and time-passed (in intervals) for the flexed-arm hang and abdominal crunch. The Command PT Representative (and additional monitors) will be posted along the 3.0 mile run course. Marines will receive their split time at the halfway mark and finishing time.

5. Logistics. Minimal special-type equipment is required to conduct a PFT.

a. Pull-Up Bar Requirements

(1) Diameter of the bar may range between 1-1¾ inches; athletic tape on the bar is authorized.

(2) The bar must be high enough to allow the tallest Marine's legs to hang straight without touching the ground, when arms are fully extended.

(3) Bench or stool may be used for shorter Marines to initially grasp the bar.

b. 3.0 mile Run Course. The run course must be measured for accuracy and set over reasonably level ground. Course selection criteria must include consideration of the following:

(1) Safety. Operational Risk Management (ORM) measures must be taken to ensure the course does not expose participants to undue risk.

(2) Performance. When possible, the course should be an "out and back" or wide loop course. The run course should not include numerous sharp turns, etc. that would force a participant to slow down excessively to remain on the course.

(3) Location. An indoor or outdoor track is permissible for the conduct of the run event. A determination as to whether the track is a yard or meter track must be made and the track measured to ensure accurate distance, prior to conducting the PFT. Run courses should not require a Marine to do more than 12 laps to cover 3.0 miles, (e.g., a course should not be shorter than 440-yard per lap). Running this event on a treadmill is not authorized.

c. Support Equipment & Personnel (Monitors)

(1) A timepiece (digital or stopwatch) that accurately measures time to the second is required for the flexed-arm hang, abdominal crunch, and 3.0 mile run.

(2) Safety equipment and personnel will be required as prescribed per local SOP.

(3) Adequate supervisory personnel to monitor each event and act as timekeepers.

d. PFT and Body Composition Worksheets. Appendix F contains the worksheet for recording PFT and body composition data. The following information will be recorded on the worksheet and be maintained by the Command PT Representative:

(1) The Marine's name, SSN, rank, age, gender, height, weight, percent body fat (if required), age 46 and Over Risk Factor Screening (if required), Physical Performance Evaluation qualifier (if required), event time(s)/total number of repetitions, event score, total score, and age-based PFT classification.

(2) Points for each event will be awarded according to the PFT Scoring Table located in appendix G.

(3) Minimum acceptable performance requirements by event (time and repetitions), and age group are listed in table 2-1 in this chapter.

(4) Minimum PFT classification scores are listed in table 2-2 in this chapter.

2101. EXECUTION OF THE PFT

1. Individual performance on each PFT event will be executed as follows:

a. Pull-up. The goal of the pull-up event is for the Marine to execute as many accurate and complete pull-ups before dropping off the bar. The procedures are:

(1) This is not a timed event.

(2) Sweatshirts will be removed during the conduct of the pull-up event in order to observe the lockout of the elbows with each repetition.

(3) Assistance to the bar with a step up, being lifted up, or jumping up is authorized. Any assistance up to the bar will not be used to continue into the first pull-up.

(4) The bar must be grasped with both palms facing either forward or to the rear.

(5) The correct starting position begins when the Marine's arms are fully extended beneath the bar, feet are free from touching the ground or any bar mounting assist, and the body is motionless.

(6) The Marine's legs may be positioned in a straight or bent position, but may not be raised above the waist.

(7) One repetition consists of raising the body with the arms until the chin is above the bar, and then lowering the body until the arms are fully extended; repeat the exercise. At no time during the execution of this event can a Marine rest his chin on the bar.

(8) The intent is to execute a vertical "dead hang" pull-up. A certain amount of inherent body movement will occur as the pull-up is executed. However, the intent is to avoid a pendulum-like motion that enhances the ability to execute the pull-up. Whipping, kicking, kipping of the body or legs, or any leg movement used to assist in the vertical progression of the pull-up is not authorized. If observed, the repetition will not count for score.

(9) A repetition will be counted when an accurate and complete pull-up is performed.

b. Flexed-Arm Hang. The goal of the flexed-arm hang event is for a Marine to hang (maintain elbow flexion) for as long as possible. The procedures are:

(1) This is a timed event.

(2) Sweatshirts will be removed during the conduct of the flexed-arm hang event in order to observe when the Marine has completely locked-out her elbows.

(3) Assistance to the bar with a step up, being lifted up, or jumping up to the start position is authorized.

(4) The bar must be grasped with both palms facing either forward or to the rear.

(5) The correct starting position begins when the Marine's arms are flexed at the elbow, the chin is held above the bar and not touching it, and the body is motionless. At no time during the execution of this event can a Marine rest her chin on the bar.

(6) Marines are authorized to drop down below the bar, however, some degree of elbow flexion must be maintained with both arms. Once a Marine's arms are fully extended or the Marine drops off the bar, the clock will stop.

c. Abdominal Crunch. The goal of the abdominal crunch event is for a Marine to execute as many proper and complete crunches within the prescribed time limit. The procedures are:

(1) 2-minute time limit.

(2) On a flat surface, Marines will lie flat on their back with shoulder blades touching the deck, knees will be bent, and both feet will be flat on the deck.

(3) The arms will be folded across the chest or rib cage with no gap existing between the arms and chest/rib cage. Both arms must remain in constant contact with chest/rib cage throughout the exercise. A single repetition consists of raising the upper body from the starting position until both forearms or elbows simultaneously touch the thighs, and then returning to the starting position with the shoulder blades touching the deck.

(4) The buttocks will remain in constant contact with the deck throughout the event. No arching of the lower back or lifting the buttocks is permitted.

(5) An assistant may be used to hold a Marine's legs or feet, at or below the knees in whatever manner that is most comfortable for the Marine. Kneeling or sitting on the Marine's feet is permitted.

(6) A repetition will be counted when an accurate and complete abdominal crunch is performed.

d. 3.0 Mile Run. The goal is for a Marine to complete the measured course as quickly as possible. The procedures are:

(1) This is a timed event.

(2) On the command to start, the two Marines monitoring the event will start their watches simultaneously when the last Marine passes the starting point. The Command PT Representative will remain at the start/finish and the other monitor will take the safety vehicle (with communication capabilities) to the halfway point. Monitors will call out the split or finishing time as appropriate, as each Marine passes.

2. The PFT scoring table is contained in appendix G.

CHAPTER 2

PHYSICAL FITNESS TEST (PFT)

SECTION 2: STANDARDS

2200. PFT STANDARDS

1. PFT Performance Requirements. To successfully pass the PFT, Marines must complete the minimum acceptable performance requirements in each event and achieve an overall combined score for their age group as shown in table 2-1. Marines must be made aware that the minimum performance (points) in each event alone, will not total the points required for a passing score. Additional points must be earned in at least one event in order to achieve a 3rd Class PFT or better, per age group. Failure to meet the minimum requirements in any event constitutes a failure of the entire test, regardless of the total number of points earned for all three events. Marines on light or limited duty will complete the two events that they are medically qualified to participate in, and will receive credit for a PPFT. Paragraph 2202 further details the PPFT requirements.

AGE	PULL-UPS/ FLEXED-ARM	ABDOMINAL CRUNCHES	3.0 MILE RUN (MIN)	TOTAL POINTS	MIN SCORE	ADDNTL' POINTS NEEDED
17-26	3/15(SEC)	50	28(m) 31(f)	105	135	30
27-39	3/15	45	29(m) 32(f)	94	110	16
40-45	3/15	45	30(m) 33(f)	88	88	0
46+	3/15	40	33(m) 36(f)	65	65	0

Table 2-1.--Minimum Acceptable Performance Requirements for PFT/PPFT Events.

2. Individual event scores for both males and females can be viewed in appendix G.

3. PFT Classification Scores. Table 2-2 shows the minimum score required, per age group, to earn each PFT classification score. Marines should be encouraged to continually strive to perform their best and not merely accept minimum performance.

	AGE GROUPS			
PFT CLASS	17-26	27-39	40-45	46+
1st	225	200	175	150
2nd	175	150	125	100
3rd	135	110	88	65

Table 2-2.--Minimum PFT Classification Scores.

4. PFT scores will be entered into the MCTFS and on FitRpts per paragraph 2203 in this chapter.

2201. ALTITUDE 3.0 MILE RUN. Marines permanently assigned or on TAD to locations at or above 4500 feet above sea level will have adjusted run times. The altitude 3.0 mile run point values are contained in appendix H. Commands are directed to provide newly joined Marines a 30-day acclimation period prior to conducting a semi-annual PFT. Marines scheduled to report to a command at altitude in June or December will complete their PFTs prior to reporting.

> CH 1 2202. PARTIAL PHYSICAL FITNESS TEST (PPFT). Marines who are medically excused from performing a particular event of the PFT as certified by a MO (e.g., light or TLD), are required to perform a PPFT. The PPFT for score requires that Marines complete two of the three PFT events. For the event not conducted, the event score from the previous semi-annual PFT will be used. Event times/repetitions will be awarded points the same way the PFT awards points; reference appendix G to view the scoring table. As with a PFT, failure to meet the minimum requirements in either event constitutes a test failure. Reference table 2-1 to view the minimum acceptable performance requirements for each event.

1. Reporting

a. The PFT and Body Composition Worksheet will indicate the event not taken and whether a light duty, TLD, or PLD status was the reason for the PPFT.

>CH 1 b. Report PPFT performance (pass/fail) for Sgt and above in accordance with reference (e).

c. LCpls and Cpls requiring a composite score will use the score (points) received from the last semi-annual PFT for the event missed.

d. PPFT scores will be entered on the Unit Diary and FitRpt per paragraph 2203.

2. Post-Light or Limited Duty PFT. Marines have the option to either keep the PPFT score taken for the current semi-annual period or to conduct a PFT (e.g., for promotion purposes) after being returned to full duty. The Marine will be given a minimum of 30 days and no more than 90 days to complete the PFT. This score will replace the PPFT taken during the waiver period. This is the only time a retest score will be used to replace a previous entry (partial) reported in MCTFS; this is not considered a remedial score.

3. Consecutive PPFTs. Marines who take PPFTs for two consecutive semi-annual periods, for the same injury or illness, will be referred to a MO and processed for a medical board. A MEB, PEB, or both will be conducted in order to determine the Marine’s fitness for duty, per reference (i).

2203. ADMINISTRATIVE PROCEDURES

1. PFT/PPFT Recording

a. Unit Diary. Only when the command declares that the PFT scheduled is for the semi-annual requirement, will scores be reported by a Unit Diary entry and entered in the MCTFS, per reference (e). Only one PFT score will be reported on the Unit Diary per semi-annual period. No retest or remedial PFT scores will be recorded on the Unit Diary. PFTs taken for inventory purposes, such as a Marine reporting into a new unit or school, or taken for screening checklist purposes, will not be reported on the Unit Diary. In addition, PFT scores conducted for unit PT or for Remedial Physical Conditioning Program (RPCP) testing may not be used (or replace a score for) the semi-annual requirement.

b. FitRpt. All PFT score reporting will be done per reference (d), paragraph 4003.8b.

(1) Semi-annual PFT scores will be recorded in section A, "Administrative Information," under "Special Information" in block 8b.

(2) Mandatory comments will be made in section I, "Directed & Additional Comments," for the following circumstances:

(a) Superior Performance. Score of 285 and above.

(b) Not Medically Qualified (NMED). Limited duty status. Refer to paragraph 2300.3 for further details.

(c) PPFT. Sgts and above that have been granted a medical excuse by a MO, and who complete and pass a PPFT will have PART entered in block 8b, section A of their FitRpt. A brief explanation of the necessity for the PPFT, (e.g., TLD or PLD must be included in section I. A failed PPFT will be reported the same as for the PFT.

(d) PFT/PPFT Failure. A PFT/PPFT failure requires that Sgts and above receive an adverse FitRpt at the end of the next reporting period. In the case where a Marine does not have a reporting period until after the next semi-annual PFT, the previous PFT failure must be recorded on that FitRpt report, regardless of the performance on the more recent one.

(e) Required but Did Not Take (RDNT). A RDNT requires Sgts and above receive an adverse FitRpt. RDNT will be recorded in block 8b, section A and an explanation must be included in section I of the FitRpt.

(3) Height and weight measurements (and percent body fat, when necessary) will be transcribed from the PFT and Body Composition Worksheet for that reporting period.

c. Proficiency/Conduct (PRO/CON) Marks. Any Marine ranked Cpl or below, will receive commensurate PRO/CON marks for superior performance, PFT failure, or RDNT per reference (g), para 4005.

d. Page 11 SRB/OQR Entries. Page 11 entries are not required for Marines who fail a PFT/PPFT or for when a PFT/PPFT is RDNT.

2. PFT/PPFT Failures

a. Promotion/Reenlistment/Special Schools. Marines may be denied promotion, reenlistment, and/or special schools until they pass a PFT and fulfill all RPCP requirements. See paragraphs 1204 and 6009.2.e of reference (j).

b. Administrative Separation. Marines may be subject to administrative separation, per chapter 6 of reference (h), paragraphs 6105 and 6206, for the following circumstances:

(1) Unsatisfactory Performance. Repetitive PFT failures, due to lack of physical conditioning, displays poor self-discipline in a Marine's responsibility to maintain fitness standards.

(2) Medical Condition. Repetitive PPFT failures due to a medical condition.

3. Summary of Administrative Actions. Table 2-3 lists the administrative actions required for Marines who fail the PFT/PPFT.

ADMINISTRATIVE ACTION	NUMBER OF FAILURES	
	PFT 1st	PFT 2nd+
Mandatory RPCP	YES	YES
Page 11 SRB/OQR Mandatory Entry	NO	NO*
Adverse FitRpt or Substandard PRO/CON Marks	YES	YES
Eligible for Promotion	NO	NO
Eligible for PCS Transfer:	YES	YES
Process for Administrative Separation	NO	YES
May Be Denied Re-enlistment	YES	YES
May Be Denied Special Schools	YES	YES
May Be Denied Recommendation for Promotion	YES	YES

Table 2-3.--Administrative Actions.

*[Page 11 entries are required when Marines have failed consecutive PFTs or PPFTs, and when process for administrative separation is anticipated. See paragraph 6206, reference (h).]

CHAPTER 2

PHYSICAL FITNESS TEST (PFT)

SECTION 3: MEDICAL CONSIDERATIONS

2300. SCOPE. When a Marine is unable to take or pass the PFT due to a physical or medical condition, then a medical evaluation by a MO must be conducted, in order to excuse the Marine from participating in a particular event or the entire test. If the condition becomes extended or considered permanent in nature, then a determination by a formal MEB, a PEB, or both must be conducted to determine the Marine's medical qualification for continued active service. Marines evaluated for or assigned to a light duty or limited duty status are still required to perform those events of the PFT that are not affected by the condition, unless otherwise specified by the MO. A Marine may not be repeatedly excused from participation in the PFT/PPFT without a MEB action taken.

1. Light Duty. The status a MO assigns to a Marine when it is determined that a medical condition exists that temporarily interferes with the performance of their duty, but the Marine is expected to return to full duty (RTFD). A MO may recommend up to 30 days light duty. A Marine that is reevaluated and not returned to a full duty status in 30 days will be placed on medical hold, and a MEB must be initiated and completed within the following 30 days. Reference paragraph 2301 for further details on the MEB.

2. TLD. The status a MO assigns to a Marine when it is determined that a medical condition exists that requires an extended period of time before they can RTFD.

a. Marines, who are on TLD and are medically exempt from two consecutive PFTs or PPFTs for the same injury or illness, will be referred to a MO and processed for a MEB, a PEB, or both to determine fitness for duty.

b. A MEB must be conducted in order to assign a Marine to a TLD status. A Marine can be assigned to a TLD status for no more than 1-8 months by a medical treatment facility (MTF).

c. A PEB must be conducted in order to grant a Marine an extension of TLD, to be assigned a PLD status, or to be processed for separation.

3. Post-Light or Limited Duty PFT. Marines removed from light or TLD status by a MO, who did not take a PFT/PPFT for the semi-annual requirement, will be administered a PFT. COs should be attentive to the fact that a Marine returned to full duty status will need to train at a moderate pace to return to the level of fitness experienced prior to the injury/illness. As a result, a Marine will be given no less than 30 days and no more than 90 days preparation after being returned to a full duty status, to complete the PFT. If the grace period crosses over into the next semi-annual interval, the PFT score will satisfy the later requirement and PFT MED (medically excused) will be entered into MCTFS for the earlier testing period.

2301. MEB. A MEB is a group of MO's at a local MTF convened for the purpose of identifying the physical and mental qualifications of referred Marines to RTFD. The MEB is charged with determining whether a Marine's limitation(s) preclude their RTFD within a reasonable period of time or at all. A MEB will evaluate and report on the diagnosis, prognosis, possible RTFD, plan for further treatment, and make medical recommendations for disposition. Reference appendix K of this Manual for further details on when a MEB is required.

1. Case Referrals. A MEB case referral will be made on Marines not returned to a full duty status after the maximum 30 days of light duty, or medically waived from completing a PFT or PPFT for two consecutive semi-annual periods for the same medical condition. A MEB may recommend:

a. Returning the Marine to full duty.

b. A period of TLD if the initial light duty period of 30 days has not rectified the problem.

c. Placing the Marine on medical hold. Medical hold is a temporary status, not to exceed 30 days, which is used to allow a MEB to convene and further evaluate a Marine's case.

d. Forwarding the case to CMC (MMSR-4) for review.

(1) The MTF must notify the Marine's command and must notify and forward a copy of the abbreviated limited duty form to CMC (MMSR-4) via naval message.

(2) Upon review, TLD may be approved, or the MEB may forward the case to the PEB for determination of fitness for continued military service.

e. Assigning the Marine to TLD. TLD is an assignment of limited duty for a specified period. Assignment to TLD can only be made after the completion of a MEB, and only if the prognosis is that the Marine can be RTFD within the specified period. The cumulative period of TLD will not exceed 16 months. The MCTFS duty limitation code for this status is "Q, LD MED BOARD" per paragraph 1206 of reference (i). The Marine's command is responsible for entering the code into MCTFS when TLD is authorized.

(1) Enlisted Assignment to TLD. A MEB recommending an initial period of TLD up to 8 months for enlisted Marines may be approved at the local MTF without approval from CMC (MMSR-4).

(2) Officer Assignment to TLD. A MEB recommending any period of TLD for Marine Officers must be submitted to CMC (MMSR-4) for review and approval.

(3) Medical Reevaluation. A MEB reassigned TLD period. The MEB at the local MTF must inform the command and CMC (MMSR-4) of the Marine's updated medical status prior to the completion of the TLD period.

(4) Assignment Extensions. Subsequent enlisted and officer assignment that extend the TLD status beyond 8 months must be submitted to CMC (MMSR-4) for review and approval.

f. Forwarding the case to a PEB for determination of fitness to continue military service. Chapter 8 of reference (h) and paragraph 2302 below further detail actions to be taken.

2302. PEB

1. The PEB acts on behalf of the SecNav to make determinations of fitness for continued military service, entitlement to benefits, disability ratings and disposition of the medical case. A PEB is required to convene for cases that exceed the maximum 16 months of limited duty or when it is not anticipated that a Marine will RTFD consequent to the initial diagnosis and prognosis. Reference appendix K of this Manual for further details on when a PEB is required.

2. PEB Findings

a. "Fit" for Continued Naval Service. To be found "Fit" to continue Naval service by the PEB, evidence must exist to demonstrate that the Marine is reasonably able to perform (without medical impairment) the duties of their billet, grade, or rank remaining on their period of service obligation to include Reserve obligation. Marines found "Fit" by the PEB are eligible for appropriate assignment. A finding of "Fit" does not preclude subsequent determinations of unsuitability for deployment, PFT participation, disqualification for special duties, TLD, or administrative action (including separation) resulting from such determination.

b. "Unfit" for Full Duty. To be found "Unfit" by the PEB to continue Naval service, evidence must exist to demonstrate that the Marine is unable to reasonably perform the duties of their billet, grade, or rank remaining on their period of service obligation to include Reserve obligation. The PEB will direct CMC (MMSR-4) to separate or retire Marines found "Unfit" to continue Naval service.

c. PLD. PLD is a continuation on active duty in a PLD status, which is approved after a Marine has been found "Unfit" by a PEB as a result of physical disability. PLD is assigned when a Marine is not expected to RTFD earlier than 16 months or it is projected that the condition cannot be corrected. A Marine can remain on active duty in a PLD status, provided the criteria (e.g., $\geq$ 18 and $\leq$ 20 years active duty) is met per reference (d). CMC (MMSR-4) may screen, review, and grant requests for PLD not meeting the mentioned criteria on a case-by-case basis.

(1) SRB/OQR Entry Requirements. When a permanent medical condition has been identified, appropriate entries will be placed in the Marine's health record and SRB/OQR. This will eliminate the necessity to repeatedly issue medical waivers. The health record entry will include MEB/PEB findings, including the date of the medical determination, diagnosis, prognosis, and restrictions imposed. The NAVMED 6100/5, Abbreviated Limited Duty Medical Board Report, will be filed on the document side of the SRB/OQR after completing the MCTFS entry requirements.

(2) MCTFS Entry. PLD is requested through the PEB and approved by CMC (MMSR-4). PLD requests will be authorized by the PEB based on the best interest of the Marine Corps and the health of the Marine. The MCTFS duty limitation code for this status is "S, RETN LD STAT" per paragraph 1206 of reference (i). Only CMC (MMSR-4) is authorized to enter or remove this code from MCTFS.

(3) FitRpt Requirement. Marines on limited duty for an entire semi-annual period will have "not medically qualified (NMED)" recorded in block 8b of section A and an explanation included in section I of the FitRpt per reference (d).

(4) A Marine assigned to PLD for a period greater than 6 months, must be reevaluated and have a complete MEB sent to the PEB no later than 4 months prior to PLD expiration date. A copy will be forwarded to the CMC (MMSR-4). This permanent condition will be reevaluated and updated at each service-related required physical evaluation (e.g., annual, reenlistment, etc.).

2303. BODY COMPOSITION STANDARDS AND LIGHT OR TLD. Marines assigned a light or TLD status or placed on medical hold for MEB/PEB screening will maintain body composition standards. Being screened for or placed on light or TLD does not absolve the Marine's responsibility to comply with Marine Corps standards. A Marine that fails to maintain body composition standards may be placed on a BCP, assigned to a modified RPCP (reflecting physical restrictions as determined by a MO), and given dietary adjustments to follow. Light or TLD Marines assigned to a BCP, who fail to successfully return to Marine Corps standards upon completion of the Program, are subject to administrative action to include a Page 11 counseling entry for unsatisfactory performance per paragraph 6105, and processed for administrative separation per reference (h).

2304. REELISTMENT/PROMOTION

1. Enlisted

a. Reenlistment. Enlisted Marines shall not be granted reenlistment while assigned to a PLD status unless the CMC grants a specific waiver. Instead, the Marine should be placed on medical hold, Convenience of the Government Medical (COFGM), until found "Fit" or "Unfit" by a PEB per paragraph 8607 of reference (h).

b. Promotion (TLD). Enlisted Marines assigned a TLD status do not lose their eligibility to be considered for promotion, due to medical conditions diagnosed by a MO that limits full duty status. Refer to paragraph 8608.8 of reference (k).

c. Promotion (PLD). Enlisted Marines who have been previously selected and are pending a fitness for duty determination by the Disability Evaluation System (DES) will be presented their certificates of appointment, and their promotions will be affected, if otherwise qualified. However, enlisted Marines assigned to a PLD status on the convening date of the appropriate selection board are not eligible for consideration for promotion.

2. Officers. If an officer is in a PLD status, delays of promotion can be initiated at the time an officer is selected by a selection board or prior to the time an officer's promotion is to be effected. Delay of an officer's promotion is warranted if the SecNav, CMC, or CO believes the officer is mentally, physically, morally, or professionally not qualified for promotion to the next higher grade. Refer to paragraph 8002.9 of reference (h) and paragraph 6003 of reference (m).

2305. PREGNANCY/POSTPARTUM. After medical determination of pregnancy, Marines will be waived from the PFT. Pregnant Marines will participate in a medically approved exercise program throughout the pregnancy and postpartum period, except when not recommended by a MO.

1. Resuming Physical Conditioning and Testing. Postpartum Marines will be required to take and pass a PFT no later than 6 months from the date that a MO has authorized her RTFD. This full duty authorization will normally coincide with the end of the 6-week (postpartum) convalescent leave period, per reference (l). Unless medically prohibited, physical conditioning should resume at this time. COs should be attentive to the fact that a Marine RTFD status will need to train at a moderate pace to return to the level of fitness she experienced prior to her pregnancy.

2. Early Termination/Miscarriage. A Marine whose pregnancy results in a miscarriage or has been terminated will be required to conform to body composition standards and pass a semi-annual PFT, in a time period determined by a MO. Just as with each pregnancy, recovery periods will vary depending on the individual's physical status at the time of termination/miscarriage. Time limits will not exceed those that are prescribed for a full term pregnancy.

3. Difficult Pregnancy or Birth/Cesarean Section/Still Birth. Although rare, these types of instances may result in a more difficult recovery. Therefore, in pregnancies where complications develop, MOs will determine on a case-by-case basis the time needed to return a Marine to full duty. Upon RTFD, the Marine will be required to conform to the body composition standards and pass the semi-annual PFT in the time period determined by a MO.

4. Lines of Communication. It is important that an open dialogue and coordination between the Marine, the MO, and the CO be maintained in order to determine any physical limitations and dietary concerns, the healing time required to return the Marine to full duty, and the Marine's overall progress.

5. Failure to RTFD. Marines not RTFD at the termination of the 6-week (postpartum) convalescent leave period, may be given an additional period of up to 30 days light duty, to allow additional recovery time. If upon termination of the 30 days light duty, the Marine is still not RTFD, then the Marine will be placed on medical hold and referred to a MEB for determination of fitness for continued military service. A Marine not RTFD status and as a result referred to a MEB will not be assigned to the BCP for failure to meet weight and body fat standards. Chapter 2, paragraph 2301 further details the MEB process and appendix L describes the Pregnancy and (inactive) BCP Assignment Process.

6. Promotion and Reenlistment

a. Pregnant Marines who did meet the reenlistment or promotion criteria prior to medical determination of pregnancy are eligible to reenlist and be promoted at any time during their pregnancy or (postpartum) convalescent period.

b. Pregnant Marines who did not meet reenlistment or promotion criteria prior to medical determination of pregnancy will not be eligible to reenlist or be promoted during the pregnancy period. Reenlistment and promotion eligibility will occur for these Marines only after they have been RTFD and have met all criteria.

2306. MEDICAL WAIVERS/EXEMPTIONS

1. Medical Waivers. A medical waiver (light or TLD) that excuses a Marine from participating in a PFT will not be considered a PFT failure. The MO will document on the medical evaluation the medical condition, the expiration date, physical exercise (based on limitations), and the PPFT events that the Marine can complete, if any, during the waiver period. Marines who are medically excused from performing a particular event of the PFT, as certified by a MO, are required to perform a PPFT. Reference paragraph 2202 of this Manual for PPFT requirements.

2. Injury During PFT. Marines injured during the execution of a semi-annual PFT (e.g., pulled muscle during the run) may be given credit for a PPFT score if two of the three events were completed prior to the injury. Or, after clearance from a MO, the Marine will be given an opportunity to run the PFT again. Only the CO and the Command PT Representative, after reviewing the facts surrounding the incident, can authorize that credit be given for a PPFT or if a Marine is given an opportunity to run the PFT at a later time. Marines completing only one event

prior to the injury will receive neither a PFT score nor PPFT credit. If evaluated and assigned a light duty status as a result of the injury, then the Marine will be required to retake the PFT 30-90 days after being RTFD status. Marines who are evaluated and not assigned a light duty status, will retake the PFT at the CO's discretion.

3. <u>Illness</u>. Marines encountering a rapid onset illness prior to the semi-annual PFT will obtain a light duty chit. Light duty chits must be submitted to the command prior to the scheduled PFT. The CO will decide when the Marine will be required to take the missed PFT.

2307. <u>AGE 46 AND OVER RISK FACTOR SCREENING</u>

>CH 1 1. Marines age 46 years and over will self-validate a "Risk Factor Screening" as a precautionary measure, 30-60 days prior to each semi-annual PFT. However, during IGMC/CGI team inspections, the inspectors may direct personnel age 46 and over to complete the Risk Factor Screening checklist if one is not on file. Marines will complete sections (A) and (B) of the Risk Factor Screening contained in appendix E, and submit the completed form to their CO for review. A medical screening by a MO, prior to PFT/PPFT participation, will only be required if the Marine:

a. Answers "yes" to any question in section (A) or (B).

b. Exceeds the body composition standards.

2. Marines requiring a follow-on medical screening must be medically cleared by a MO prior to participating in the PFT. Supplemental SF600 forms will be attached to the Screening if additional MO remarks are required. Independent Duty Corpsmen are not authorized to issue medical clearance to Marines who answered "yes" to any risk factor screening question. Marines who are not cleared by a MO to participate in the semi-annual PFT are restricted from participation during the current semi-annual period, either voluntarily or involuntarily. Such restriction only applies to the semi-annual period during which the MO determination was made.

3. Marines will participate in the PFT only if cleared by a MO after the referral, and no further medical evaluation is required. The signed and completed Risk Factor Screening (original) and SF600 documents will be filed in their health records, and a copy of the screening will be given to the CO. The Command PT Representative will annotate the Marine has completed the screening (been cleared for participation), on the PFT and Body Composition Worksheet.

4. Marines having a current physical examination or Navy Annual Health Evaluation (BUMEDINST 6120.26) will not be required to complete the semi-annual Risk Factor Screening or receive additional medical clearance to participate in the PFT if:

a. Examination or evaluation was conducted within the last 6 months from the date of the scheduled PFT.

b. There were no physical restrictions or limitations imposed by the MO.

c. Their state of health has not since changed.

CHAPTER 2

PHYSICAL FITNESS TEST (PFT)

SECTION 4: REMEDIAL PROCEDURES

2400. RPCP

1. The command directed RPCP is designed to improve a Marine's physical fitness and body composition, through command-structured physical conditioning, guidance, and supervision as contained in appendix C.

2. Assignment. Marines will be placed on a RPCP for any of the following:

a. Failing a semi-annual PFT.

b. Marines assigned to the BCP.

c. Marines required to take a semi-annual PFT but fail to do so ("required did not take, RDNT").

d. Marines who consistently demonstrate poor physical performance in MOS related tasks.

e. Female Marines assigned to the BCP prior to pregnancy and RTFD status will resume active participation in the RPCP. The level of program participation may require MO input, if physical limitations are identified.

3. Program Requirements. RPCP requirements:

a. Minimum of 30 days in the Program, regardless of change in ability.

b. Take and pass a PFT at completion of the initial 30-day period.

c. Marines assigned to the BCP will remain in the RPCP for the duration of the BCP assignment.

d. No formal administrative action is required for assignment to RPCP (e.g., no requirement for a SRB/OQR, Unit Diary, or MCTFS entry).

CHAPTER 3

BODY COMPOSITION PROGRAM (BCP)

CHAPTER 3

BODY COMPOSITION PROGRAM (BCP)

3000. SCOPE. It is essential that every Marine maintain the established standards of health and physical fitness. With age, it is not uncommon for the average American to experience a decrease in physical activity, a change in metabolic rate, with unchanged caloric intake. As a result, body fat and weight may increase, having a negative impact on one's health and level of fitness. Medical authorities state that excess body fat can be associated with high blood pressure, high blood cholesterol, diabetes, cancer, cardiovascular disease, and similar health risks. In extreme climates and physically demanding environments, excess weight and body fat can also impede physical performance and stamina, which is of particular importance to the Marine and the mission.

3001. OBJECTIVES. The objectives of the Marine Corps' BCP are:

1. To establish healthy weight and body fat standards that ensure all Marines are physically capable to perform their duties.

2. To contribute to the health and well being of every Marine by continuously monitoring weight, body fat, diet/nutrition, and physical fitness conditioning.

3. To motivate all Marines to set the example by maintaining the established body composition standards.

4. To ensure those Marines who do not meet established standards are counseled accordingly, and given the opportunity and proper guidance to achieve the standards through the BCP and Remedial Physical Conditioning Program (RPCP).

3002. RESPONSIBILITIES

1. Individual Marine. It is every Marine's responsibility to maintain the Marine Corps' body composition standards, or if not in adherence with standards, to take appropriate action to return to them in a timely manner. Marines on light or limited duty, whose medical condition precludes them from participating in specific activities, will be expected to participate in conditioning alternatives and dietary adjustments, in order to maintain these standards.

2. CO. It is the CO's responsibility to ensure each Marine receives a semi-annual Body Composition Evaluation, and to closely monitor Marines assigned to the command's BCP and RPCP. COs will:

a. Continually monitor all Marines within the command to ensure adherence to established body composition standards.

b. Identify Marines who exceed the weight and body fat standards by conducting periodic Body Composition Evaluations.

c. Offer educational programs, events, and other motivational means to encourage Marines to achieve and maintain appropriate body composition standards. Such programs could include, but are not limited to: body composition seminars, health and fitness fairs, and individual consultations conducted by qualified dieticians, medical authorities, and Semper Fit personnel. Spouses should be encouraged to participate in these programs with the assigned Marine. Reference the Semper Fit website (www.usmc-mccs.org) for additional information and services.

d. Ensure the leadership understands that Marines may not have the education and experience needed to attain and maintain proper health and fitness, and that lifestyle changes may take time. However, if given the proper tools (e.g., education, supervision, and positive reinforcement), leaders can assist Marines in achieving and maintaining a healthy and fit lifestyle more quickly.

3. Medical Officer (MO). The MO's responsibility is to evaluate the Marine's body composition, overall health, diet, and physical ability to participate in the BCP and RPCP. The CO should receive the following from the MO:

a. Determination of whether the weight or body fat increase is due to an underlying cause or associated disease.

b. The Marine's current duty status (e.g., full, light, or limited duty).

c. The recommended dietary plan and weight and body fat reduction goals.

d. The physical conditioning restrictions or limitations imposed.

e. The date the Marine is expected to return to full duty (RTFD) status.

f. Any other pertinent information that pertains to the Marine's individual case.

CHAPTER 3

BODY COMPOSITION PROGRAM (BCP)

SECTION 1: STANDARDS

3100. BODY COMPOSITION STANDARDS. The Marine Corps' weight and body fat standards are health and performance based, and not based on appearance. Marines are considered not within these standards when their body weight and body fat exceed the maximum limits as contained in appendix I. If tested and the Marine's percent body fat exceeds the maximum limit (18% for males, 26% for females), then the CO will evaluate the Marine's physical performance (e.g., current semi-annual Physical Fitness Test (PFT)) using the Physical Performance Evaluation in paragraph 3102. If the Marine does not meet the Physical Performance Evaluation criteria, then the CO will have the Marine evaluated by a MO and assigned to a BCP - formerly known as the "Weight Control Program." Marines assigned to the BCP will receive assistance in reducing body weight and in particular body fat, in order to attain and maintain a more healthy physical fitness state. The BCP assignment process and endorsements are contained in paragraph 3201 and appendix D of this Manual.

3101. BODY COMPOSITION EVALUATION PROCEDURES

>CH 1 1. Body Composition Evaluation (BCE). The BCE is a semi-annual (January-June and July-December) requirement (annual for Reserves) designed to ensure Marines are within HT/WT or BF standards. However, commanders may direct BCE's for a unit or individuals as often as deemed necessary. BCE's may be conducted on the same day as the PFT. IGMC/CGI teams may also conduct unannounced BCE's in support of inspections.

a. Weigh-Ins. Regardless of duty status (full/light/limited duty), all Marines are required (at a minimum) to complete a semi-annual (annual for Reserves) Body Composition Evaluation as prescribed by the unit training schedule. The Command Physical Training (PT) Representative and his/her designate are the **only** unit personnel authorized to conduct Body Composition Evaluations, and will record the results on the PFT and Body Composition Worksheet in appendix F.

(1) Uniform. Uniform for the Evaluation is the USMC established green-on-green T-shirt and shorts, and socks. No other uniform or PT attire is permitted. Shoes will not be worn.

(2) Height Measurement. Height will be measured using a standard, non-stretching (cloth or fiberglass) tape measure that is secured to a wall and flush with the deck; weight scales that have a height measuring device will not be used. The Marine will stand with their back against the wall, heels flat on the deck, shoulders back, with arms to the side in a relaxed manner, and head straight forward. A ruler will be used to mark the Marine's exact height.

(a) Measurement will be recorded to the nearest inch.

(b) If the height fraction is less than 1/2 inch, round down to the nearest inch. If height fraction is 1/2 inch or more, round up.

(3) Weight Measurement. Weight will be measured on a calibrated balance beam scale in the required PT uniform, without running shoes. A one-pound reduction will be granted for the PT uniform; however, no other weight reductions are authorized.

(a) Measurement will be recorded to the nearest pound.

(b) If the weight fraction is 1/2 pound or less, round down to the nearest pound. If more than 1/2 pound, round up.

(4) Refer to section I of appendix I to review the height and weight chart. If a Marine is not within height and weight standards, then the Marine will be measured for body fat immediately (same day). Body fat estimation is only authorized for Marines who exceed their maximum weight standard. The height and weight measurement conducted during this portion of the Body Composition Evaluation is what will be reported on the FitRpt.

b. Body Fat Estimation. Body fat will be estimated using the circumference-based method with one set of measurements for each gender (Males: height, neck circumference, and abdominal circumference at the navel. Females: height, neck circumference, waist circumference at the thinnest portion of the abdomen, and hips). No substitute methods of assessment are permitted. This DoD directed method of body fat estimation has been carefully evaluated for applicability to service members and represents the best approach that can be applied with minimal error (+/- 3 to 4%). This method is also valid because of the emphasis on abdominal circumference, the site of human body fat deposition which is most strongly associated with health risks, and which corresponds to other military goals including appropriate appearance and healthy exercise habits. Body fat measurements will be taken in the following manner:

(1) Marines will report for body fat estimation in the required PT uniform. The Command PT Representative and his/her designate will conduct the measurements.

(2) The tape measure used should be made of non-stretchable material, preferably fiberglass; cloth or steel tapes are unacceptable. The tape measure width should be 1/4 - 3/8 inch. The tape measure should be calibrated (i.e., compared with a yardstick or a metal ruler to ensure validity). This is done by aligning the fiberglass tape measure with the quarter inch markings on the ruler. The markings should match those on the ruler; if not, do not use that tape measure.

(3) All measurements will be taken on bare skin with the only exception being the female hip measurement.

(4) For use in the body fat estimation formula, Marines will have their height re-measured and recorded to the nearest 1/2-inch.

(a) If the measured height fraction is greater than an inch, round up to the nearest 1/2-inch.

(b) If the measured height fraction is greater than 1/2-inch, round up to the next whole inch.

(c) If the measured height fraction is at 1/2 inch, use that measurement in the body fat estimation equation.

(d) The height measured at the initial semi-annual (annual for Reserves) height/weight screen is what will be reported on the FitRpt, not the height used for the body fat estimation formula.

(5) Without causing indentation, the tape will be applied to the skin with sufficient tension to hold it in place as the measurement is being read.

(6) Circumference measurements will be taken three times to ensure accuracy. Each set of measurements will be completed sequentially to discourage assumption of repeated measurement readings. For males, complete one set of abdomen and neck measurements, **not**, for example, three abdominal circumferences followed by three neck circumferences. Continue the process by measuring the abdomen and neck in series until three sets of measurements are completed. For females, complete one set of waist, hip, and neck, measurements, **not**, for example, three waist measurements followed by three hip measurements. Continue the process by measuring the waist, hip, neck series three sets of measurements are completed. The lower of the three body fat measurements will be recorded on the PFT/Body Composition Worksheet in appendix F.

(7) Procedures for Body Fat Estimation for Males:

(a) Measure the neck circumference by placing the edge of the tape measure flush with the bottom of the larynx and perpendicular to the long axis of the neck. The Marine should look straight ahead during the measurement, with shoulders down (not hunched). For neck measurements in excess of the whole inch, round the neck measurement up to the nearest 1/2-inch and record (e.g., round up 16 1/4 inches to 16 1/2 inches).

(b) Measure abdominal circumference against the skin at the navel, level and parallel to the deck. Arms are at the sides. Take measurement at the end of the Marine's normal, relaxed exhalation. Round the abdominal measurement down to the nearest 1/2-inch and record (e.g., round down 34 3/4 to 34 1/2 inches).

(c) Determine percent body fat by subtracting the neck from the abdominal measurement and comparing this value against the height measurement, (e.g., abdominal - neck = circumference value). Refer to the chart "Percent Body Fat Estimation for Males" in section II of appendix I.

(8) Procedures for Body Fat Estimation for Females:

(a) Measure the neck circumference by placing the edge of the tape measure flush with the bottom of the larynx and perpendicular to the long axis of the neck. The Marine should look straight ahead during the measurement, with shoulders down (not hunched). To establish neck measurements in excess of the whole inch, round the neck measurement up to the nearest 1/2-inch and record (e.g., round up 13 3/8 inches to 13 1/2 inches).

(b) Measure the natural waist circumference against the skin at the point of minimal abdominal circumference, usually located about halfway between the navel and the lower end of the sternum (breastbone). When this site is not easily observed, take several measurements at probable sites and use the smallest value. Ensure the tape is level and parallel to the deck. Arms are at the sides. Take measurements at the end of a normal, relaxed exhalation. Round natural waist measurement down to the nearest 1/2-inch and record (e.g., round down 28 5/8 inches to 28 1/2 inches).

(c) Measure the hip circumference while facing the Marine's right side by placing the tape around the hips so that it passes over the greatest protrusion of the buttocks as viewed from the side. Ensure the tape is level and parallel to the deck. Apply sufficient tension on the tape to minimize the effect of clothing. Round the hip measurement down to the nearest 1/2-inch and record (e.g., round down 38 3/8 inches to 38 inches).

(d) Determine percent body fat by adding the waist and the hip measurements, subtracting the neck measurement, and comparing values against the Marine's height measurement (e.g., waist + hip - neck = circumference value). Refer to the chart "Percent Body Fat Estimation for Females" in section II of appendix I.

(e) Marines who meet the body fat standards, 18% for males and 26% for females, will be considered within the Marine Corps' body composition standards. Marines who exceed the body fat standards will be further evaluated on criteria set forth in paragraph 3102 below.

3102. PHYSICAL PERFORMANCE EVALUATION. There is an integral relationship between body composition and physical performance that warrants consideration when evaluating the "total" Marine. For instance, there are Marines who exceed weight and body fat standards, yet perform exceptionally well on the semi-annual PFT and combat fitness tests. These Marines demonstrate both exceptional cardiovascular endurance and upper body strength, with a body composition that may be identified as slightly exceeding the standards, but overall could be viewed as maintaining a reasonable body weight. In such cases, the Marine Corps' view on physical fitness as "essential to the day-to-day effectiveness and combat readiness" should be the overriding factor in these circumstances. The Physical Performance Evaluation not only takes body composition and physical fitness performance into consideration, it also takes into account the acknowledged variance in the body fat estimation technique ("taping" method). The Physical Performance Evaluation accommodates for the acknowledged margin of error in body fat estimation and testing, normally a variance of 3 to 4%. Reference (f) establishes a requirement to make such an accommodation in view of this variance. In essence, the Physical Performance Evaluation assesses the "total" Marine (e.g., body composition and recent semi-annual PFT performance as well as acknowledges the margin of error in body fat testing).

1. The Physical Performance Evaluation requires a Marine to meet the following criteria:

>CH 1 a. A 1st Class PFT taken a maximum of 90 days before or 30 days after the body composition evaluation.

b. Body fat estimation does not exceed the standard by more than 4%, (≤22% for males, ≤30% for females).

2. Marines exceeding 22 and 30% respectively or score a 2nd Class PFT or lower will not be eligible for the Physical Performance Evaluation under any circumstance. These Marines will automatically be processed for a BCP and RPCP assignment.

3. Partial Physical Fitness Tests (PPFT) will not be considered. Marines must be fit to take a PFT and score a 1st Class to be considered for the Physical Performance Evaluation.

4. No adverse administrative action will be taken for Marines meeting the Physical Performance Evaluation criteria. The Body Fat estimation percentage will still be reported on the FitRpt. The following is a directed comment in section I on the FitRpt meeting the Physical Performance Evaluation criteria:

> "MRO meets the Physical Performance Evaluation criteria in MCO P6100.12 and is within standards."

5. Marines meeting the Physical Performance Evaluation criteria will only be exempt from administrative action for a particular 6-month (semi-annual) period (or annual for Reserves). Marines will be reevaluated at the next semi-annual PFT (annual for Reserves) and Body Composition Evaluation per paragraph 2100.

6. Physical Performance Evaluation examples:

a. Example 1: A male Marine, who exceeds his maximum allowable weight, is measured at 21% body fat, and scores a current 1st Class PFT has met the Physical Performance Evaluation criteria and would not be assigned to the BCP.

b. Example 2: A male Marine, who exceeds his maximum allowable weight, scores a current 1st Class PFT, but is measured at 24% body fat, does not meet the Physical Performance Evaluation body fat criteria, and therefore would be assigned to the BCP.

c. Example 3: A female Marine who exceeds her maximum allowable weight, is measured at 32% body fat, and scores a current 2nd Class PFT, does not meet the Physical Performance Evaluation body fat and PFT criteria, and therefore would be assigned to the BCP.

3103. PREGNANCY/POSTPARTUM BODY COMPOSITION REQUIREMENTS. Marines will be exempt from weight and body fat standards during pregnancy, but will maintain active participation in the Physical Conditioning Program (PCP) unless otherwise prescribed by her MO. Marines in a postpartum status will be required to meet weight and body fat standards no later than 6 months following a MO's authorization to RTFD. This authorization normally coincides with the completion of the 6-week (postpartum) convalescence leave period, per reference (l).

1. Miscarriage/Stillbirth. A Marine whose pregnancy terminates prematurely or that results in a stillbirth will meet weight and body fat standards within a time period determined by a MO. Due to the varying disposition of this medical situation, it is important that an open dialogue exists between the Marine, the MO and her CO to determine the time needed to recover and conform to body composition standards. The recovery period in this case, however, will not exceed the 6 months authorized for full-term pregnancies.

2. BCP Assignment. Marines who become pregnant while assigned to the BCP will remain on the Program in an "inactive" status.

a. The only Unit Diary entry required when a Marine becomes pregnant while assigned to a BCP is to report MCTFS Duty Limitation Code of "N, PREGNANCY" per paragraph 1206 of reference (i). An advisory message will be generated on the unit's Diary Feedback Report (DFR) as follows: "MARINE PLACED IN INACTIVE WT CNTL STATUS." All further advisories concerning the Marine's status in the BCP will be suppressed for the duration of the pregnancy and the medical related periods of limited duty immediately following the pregnancy.

b. Following the MO's medical evaluation, the Marine will be returned to either a full duty status, or reported to a non-medical related limited duty status. The Marine will then resume active participation in the BCP. All Marines will be assigned for a 6-month period, regardless of time spent on the Program prior to pregnancy. The unit will receive the following advisory message on their DFR: "MARINE RETURNED TO ACTIVE WT CNTL STATUS FOR SIX MONTHS."

3. Promotion Eligibility. Marines will not be eligible for promotion if they are in an "inactive" status for a BCP assignment per paragraphs 1204.3n and 6009.2e of reference (k). Only when the Marine returns to the weight and body fat standards and meets all other promotion requirements will she be promoted. Reference paragraph 2305 of this Manual for further details.

CHAPTER 3

BODY COMPOSITION PROGRAM (BCP)

SECTION 2: BCP PROCEDURES

3200. BODY COMPOSITION MEDICAL EVALUATION

1. Marines not within the body composition standards and do not meet the criteria for Physical Performance Evaluation will be referred to a MO to be evaluated. A medical evaluation and clearance, and all the Program endorsements contained in appendix D must be completed before a Marine can be assigned to a BCP. During the MO's evaluation, a determination must be made on whether or not a Marine's increase in weight and body fat is due to an underlying cause or associated disease, as follows:

>CH 1 a. Underlying Cause Defined. The following conditions may be considered as an underlying cause or associated disease which exempts a Marine (temporary basis) from meeting height/weight (HT/WT) or bodyfat (BF) standards and not being assigned to a BCP:

(1) Metabolic conditions, such as hypothyroidism, diagnosed by a medical officer (MO). The temporary exclusion will expire within 6 months of the metabolic condition being normalized through treatment or a limited duty/permanent disability board will be initiated.

(2) Medical conditions which require administration of medications that directly result in temporary weight gain.

>CH 1 b. Neuromusculoskeletal (NMS) Conditions. NMS conditions (e.g., sprained ankle, low back strain) that cause Marines to be placed in a not fit for full duty status may not be considered as an underlying cause or associated disease which exempts a Marine (temporary basis) from meeting HT/WT or BF standards. A NMS condition/not fit for full duty status determination also does not preclude assignment to a BCP. Assignment to a BCP for Marines in a not fit for full duty status secondary to a NMS condition will include referral from a MO or appropriately credentialed health care provider (ACHCP) to an appropriate agency for nutritional guidance and commencement of an alternative conditioning program.

>CH 1 c. Out of Standards Without Underlying Cause. When receiving the initial evaluation by the MO it is determined that no underlying cause or associated disease is the reason for the Marine's weight and body fat gain, then the Marine will continue forward with the BCP assignment process. The following medical recommendations will be made per an individual Marine's case:

(1) Recommend a realistic weight loss (e.g., pounds per week) and percent body fat reduction time line.

(2) Recommend caloric intake reduction and nutritional guidance (dietary measures).

(3) Recommend alternative physical conditioning exercises based on physical limitations (if necessary) for Marines on light or limited duty.

>CH 1 d. Out of Standards With an Underlying Cause. When receiving their initial evaluation by the MO, it is determined that an underlying or associated disease is the reason for the Marine's weight and body fat gain, then it must be thoroughly

annotated in the Marine's Health Record and on the Program's First Endorsement. The following courses of action may be taken per individual Marine's case:

(1) Treatment to alleviate the medical condition and return of the Marine to the command.

(2) Hospitalization for obesity treatment at an Alcohol Rehabilitation Center (ARC). Criteria for ARC Residential Obesity Program (ROP) are contained in paragraph 3202 in this Manual.

(3) Diagnosed as untreatable or permanent in nature, case forwarded to a PEB for determination of fitness to continue military service.

(4) RTFD, assignment to BCP and RPCP.

3201. <u>BCP ASSIGNMENT</u>. The following actions will be taken by the command when a Marine is being screened for a BCP assignment. Marines assigned to the BCP will be required to participate in the Program for a minimum of 6 months.

1. Preliminary Medical Evaluation. The CO will complete the BCP Evaluation Letter (appendix D, enclosure 1) requesting a preliminary medical evaluation. This Letter and the First Endorsement (appendix D, enclosure 2) will be forwarded to the MO at the local Medical Treatment Facility (MTF). The Marine will report directly to the MTF to be medically evaluated by the MO.

2. First Endorsement. The MO will complete the First Endorsement (appendix D, enclosure 2) and return the Letter to the CO.

a. The MO will attach the following information to the Endorsement:

(1) Determination of whether the weight and body fat increase is due to an underlying cause or associated disease;

(2) Recommended weight and body fat reduction goals;

(3) Recommended dietary measures;

(4) Physical limitations or restrictions imposed; and

(5) Medical clearance to participate in the Command's BCP and RPCP.

b. If the Marine is diagnosed with an underlying cause or associated disease, then the MO may recommend that the Marine not be assigned to a BCP until the Marine's condition is treated, or may recommend that the Marine be processed for medical separation for untreatable conditions. This diagnosis must be thoroughly annotated in the Marine's Health Record and on the Program's First Endorsement. Reference paragraph 3200 in this Manual for further details regarding "underlying cause and/or associated disease."

c. The signature block on the Endorsement will clearly state the name, grade, Branch of Service, medical title, and organization (as appropriate) of the MO who diagnosed the cause of the Marine's deficiency.

3. Second Endorsement. Upon reviewing the MO's diagnosis and recommendations, the CO will complete the Second Endorsement (appendix D, enclosure 3). The CO will counsel the Marine on the deficiencies, corrective actions required, and information regarding the BCP and RPCP, and will have a 6105 counseling entry recorded on the Marine's SRB/OQR's Page 11 per reference (h).

4. Third Endorsement. Upon completion of the counseling, the Marine will sign and date the Third Endorsement (appendix D, enclosure 4) acknowledging the deficiencies, corrective actions required, and requirements for the 6-month BCP and RPCP assignments.

5. Admin Section Responsibilities. The CO will direct the Admin Section to:

a. On receipt of the Third Endorsement (appendix D, enclosure 4):

(1) Enter the BCP assignment on the Unit Diary and record the Unit Diary number and date on the bottom of the Third Endorsement. Forward this Endorsement to the Command PT Representative to be maintained in the Marine's BCP file.

NOTE: A BCP assignment will not be run on the Unit Diary until the Marine has been evaluated and cleared for participation by a MO, the appropriate Program endorsements have been completed, and the Marine has been properly counseled by the CO.

(2) Use the following code/statement to report assignment to BCP in MCTFS: TTC 337 000 TO WT CNTL.

b. When directed, record 6105 counseling entry on SRB/OQR Page 11.

c. When directed, submit a request for Marine Corps Institute (MCI) Course 3316, "Basic Nutrition."

d. Notify the Command PT Representative when the Admin Section has received a Diary Feedback Report (DFR) that identifies a Marine's BCP assignment has expired.

e. When directed, make an entry on the Unit Diary extending the Marine's initial (first) BCP assignment and enter the following code/statement per reference (e) into MCTFS: TTC 337 001 EXTEND WT CNTL. Record the Unit Diary number and date on the Sixth Endorsement (appendix D, enclosure 7) and forward it to the Command PT Representative to be maintained in the Marine's BCP file.

f. When directed, remove the Marine from the BCP, making an entry on the Unit Diary and the following code/statement per enclosure (e) into MCTFS: TTC 339 000 FR WT CNTL. Record the Unit Diary number and date on the Sixth Endorsement (appendix D, enclosure 7) and return it to the Command PT Representative to be maintained in the Marine's BCP file.

6. Command PT Representative Responsibilities. The CO will direct the Command PT Representative to:

a. On receipt of the Third Endorsement (appendix D, enclosure 4) from the Admin Section, create a BCP file for the Marine. The BCP file will contain all Marine's Program Endorsements, medical documentation, counseling, body composition testing results, training log, etc.

b. Immediately initiate the 6-month RPCP for a Marine medically qualified. A Marine in a light or limited duty status will be required to participate in an alternative RPCP, based on the physical restrictions and limitations imposed by the MO.

c. Conduct diet and nutrition classes for a Marine or set up appointments with the fitness and medical professionals at the local Semper Fit facilities or MTFs.

d. Order (via Admin Section) MCI Course 3316, "Basic Nutrition," for the Marine (whether Officer or Enlisted), and direct the Marine to complete the Course prior to BCP assignment expiration.

e. Take weight and body fat measurements once every 2 weeks (once a month SMCR), and document results in the Marine's BCP file.

f. Document in the Marine's BCP file the counseling conducted for those Marines who have displayed neither weight loss nor body fat reduction, and if it is a result of the Marine not adhering to the regimen prescribed by the MO and the command.

(1) A Marine not making satisfactory progress during the BCP assignment will receive a 6105 counseling entry on the SRB/OQR's Page 11 for failure to make satisfactory progress. This Page 11 entry will be given for unsatisfactory performance per paragraph 6206 in MCO P1900.16F, and is required for administrative separation.

(2) A Marine who makes progress, but fails to meet the weight and body fat standards at completion of a BCP assignment will receive a 6105 counseling entry on the SRB/OQR's page 11. This Page 11 entry will be given for failing to meet body composition standards per paragraph 6215 in MCO P1900.16F, and is required for administrative separation.

g. Maintain and update accordingly, the BCP file, (e.g., counseling, body composition test results, and RPCP training log entries).

>CH 1 h. Sgts and above who are assigned to the BCP will be administered an adverse FitRpt per reference (d). Cpls and below will receive commensurate PRO/CON marks per reference (g).

7. Fourth Endorsement. After a Marine has participated in the BCP for 4 months, the CO will complete the Fourth Endorsement (appendix D, enclosure 5) and will forward this endorsement and the Fifth Endorsement (appendix D, enclosure 6) to the MO at the local MTF. The Marine will report to the MTF to be reevaluated by the MO. (If at any time the Marine is failing to lose weight or body fat, the Marine should be evaluated earlier, in order to be reexamined for a possible underlying or associated disease.)

8. Fifth Endorsement. The MO will reevaluate the Marine, complete the Fifth Endorsement (appendix D, enclosure 6), and forward the endorsement to the CO with one of the following diagnoses/recommendations:

a. Underlying Cause or Associated Disease. The Marine has been diagnosed as having an underlying or associated disease, and the Marine will:

(1) Be placed in an "inactive status" (duty limit "N") for the BCP assignment until the Marine's condition is treated; or

(2) Be recommended for (medical) separation for untreatable conditions.

b. Reference paragraph 3200 for further guidance on cases involving an underlying cause or associated disease.

c. No Underlying Cause or Associated Disease and Making Progress. The Marine has been diagnosed as not having an underlying cause or associated disease and is making satisfactory progress towards weight and body fat reduction goals. Recommend that participation in the BCP (and RPCP) continue until assignment expires.

d. No Underlying Cause or Associated Disease - Not Making Progress. The Marine has been diagnosed as not having an underlying cause or associated disease and is not making satisfactory progress towards weight and body fat reduction goals. Recommend that participation in the BCP (and RPCP) continue until assignment expires.

(1) A Marine may not receive an extension if lack of satisfactory progress is a result of not adhering to the regimen prescribed by the MO and the command.

(2) A 6105 counseling entry on the Marine's SRB/OQR's Page 11 will be administered at this time (fourth month) for failure to make satisfactory progress. See paragraph 6206 in MCO P1900.16F.

e. Returned to Weight or Body Fat Standards. The Marine has met weight and/or body fat reductions goals and is within body composition standards. Recommended that participation in the BCP (and RPCP) continue until assignment expires.

9. Expiration of Program Assignment. The Admin Section will receive a DFR when the BCP assignment has expired. One of the following actions will be taken:

a. A Marine has met and maintained the weight and body fat standards and will be officially removed from the BCP (and RPCP); see paragraph 13 in this section for Program removal instructions.

b. A Marine has failed to return to the weight and body fat standards, but has made satisfactory progress. Refer to paragraph 11 below for further guidance on Program (first assignment) extensions.

c. A Marine has failed to return to the weight and body fat standards and has not made satisfactory progress. CO will administer a 6105 counseling entry on the Marine's SRB/OQR's Page 11, per paragraph 6206 or 6215 in MCO P1900.16F respectively, and will process the Marine for administrative separation.

10. Extensions to First Assignment. A Marine making satisfactory progress and reasonably close to meeting their weight and body fat standards at the end of the 6-month BCP assignment may be granted a Program extension.

a. CO can grant a one-time extension of up to 6 months to first assignments only. An extension to a first assignment does not constitute a second assignment.

b. A Marine granted an extension will continue participation in the BCP and RPCP until the extension period expires. A Marine granted an extension will be counseled pursuant to paragraph 6105 of reference (h) with standard language in subparagraph (1) below and have the following code statement entered into MCTFS: TTC 337 001 EXTEND WT CNTL. Counseling pursuant to paragraph 6105 of reference (h) for Marines not making significant progress while on their first assignment or the extension to their first assignment to the BCP will occur after the initial 60-day probation period and, at a minimum, at the fourth month of the BCP assignment when the Marine is reevaluated by the MO and diagnosed as not having an underlying cause or disease. Subparagraph (2) provides the recommended standard language for this counseling.

(1) "Counseled this date concerning your extension on the Marine Corps BCP. You have made reasonable progress but have failed to reach your required body composition while on your first assignment to this Program. Therefore, you have been granted a one-time extension of 6 months to meet the body composition standards set forth in MCO P6100.12. You are advised that failure to take corrective action and meet your body composition standards or weight/ body fat reduction goals, as prescribed by the medical officer, while on this 6 month extension will result in administrative separation from the USMC for either weight control failure or unsatisfactory performance per paragraphs 6206 and 6215 of MCO P1900.16E. Specific recommendations for corrective action are: Meet your weight/body fat reduction goals by losing the prescribed __ lbs and __ percent body fat per month for 6 months and actively participate in your diet and exercise program. Assistance is available through your chain of command. I was advised that within 5 working days after acknowledgment of this entry a written rebuttal could be submitted and that such a rebuttal will be filed on the document side of my SRB/OQR. I choose (to)(not to) make such a statement."

(2) "Counseled this date concerning your unsatisfactory performance while assigned to the Marine Corps BCP. Due to insufficient effort, you have not met your weight/body fat reduction goals as prescribed by the medical officer. You are advised that failure to take corrective action and reach your required body composition will result in administrative separation from the USMC for either weight control failure or unsatisfactory performance per paragraphs 6206 and 6215 of MCO P1900.16E. Specific recommendations for corrective action are: Meet your weight/body fat reduction goal and actively participate in your diet and exercise

program. Assistance is available through your chain of command. I am advised that within 5 working days after acknowledgment of this entry, a written rebuttal could be submitted and that such a rebuttal will be filed on the document side of my SRB/OQR. I choose (to)(not to) make such a statement."

>CH 1 c. Unlike the mandatory 6 month time period associated with the initial assignment to a BCP, Marines on an extension to a first assignment will be removed from the BCP when they return to HT/WT or BF standards; see paragraph 13 in this section for program removal instructions. A Marine not meeting standards at the end of the extension will be processed for administrative separation for failure to meet body composition standards per paragraph 6215 in MCO P1900.16F. No additional extensions can be granted.

>CH 1 11. Second Assignment. A Marine who has been assigned to and officially removed from the BCP at any time in their career, and fails to meet the weight and body fat standards a second time, may receive a one-time second assignment to the BCP (and RPCP). On a second assignment, Marines will be given 6 months with no extensions to conform to standards and will be counseled pursuant to paragraph 6105 of MCO P1900.16F with the recommended standard language. Unlike the mandatory 6 month time period associated with an initial assignment to a BCP, Marines on a second assignment to a BCP will be removed from the BCP when they return to HT/WT or BF standards. A Marine will be assigned to the Program as per paragraphs 2 through 10 of this section. A Marine successfully meeting the weight and body fat standards at the completion of the second assignment will be officially removed from the BCP (and RPCP) as stated in paragraph 13 in this Section. A Marine not meeting the standards when the second assignment expires will receive a 6105 counseling entry on the their SRB/OQR's Page 11 per paragraph 6206 or 6215 in MCO P1900.16F respectively, and will be processed for administrative separation. Extensions to a second assignment are not authorized.

> "Counseled this date concerning your reassignment to the Marine Corps BCP. Specifically, you failed to properly maintain your body composition standards as required by MCO P6100.12 for a second time. Your actions are unsatisfactory and not in accordance with the high state of readiness required by the USMC. You are advised that this subsequent assignment is for a 6-month period. Specific recommendations for corrective action are: Meet your weight/body fat reduction goals by losing the prescribed __ lbs and __ percent body fat per month and actively participate in your diet and exercise program. Assistance is available through your chain of command. You are advised that failure to take corrective action and reach the Marine Corps body composition standard will result in administrative separation from the USMC for either weight control failure or unsatisfactory performance per paragraphs 6206 and 6215 of MCO P1900.16E. Additionally, you are advised that MCO P6100.12 precludes assignment to the BPC for a third time in a Marine's career and that any subsequent failure to meet prescribed weight/body fat standards during your career will result in administrative separation. I am advised that within 5 working days after acknowledgment of this entry, a written rebuttal could be submitted and that such a rebuttal will be filed on the document side of my SRB/OQR. I choose (to)(not to) make such a statement."

12. Follow On Assignments. A Marine, assigned to the BCP on two separate occasions (e.g., first and second assignments), that fail to meet weight and body fat standards a third time will automatically be processed for administrative separation per paragraph 6215 of MCO P1900.16F. In such cases, the Marine will still be required to participate in the BCP and RPCP until discharged; no formal assignment is required. A prior counseling entry pursuant to paragraph 6105 of MCO P1900.16F is required to separate a Marine for either weight control failure or unsatisfactory performance per paragraphs 6206 and 6215 of MCO P1900.16F.

13. Program Removal. A Marine will not be removed from the BCP (and RPCP) assignment prior to the Program expiration date. A Marine will continue to participate in the BCP and RPCP as prescribed, regardless of progress, in order to ensure healthy weight and body fat loss techniques were employed and the weight and body fat loss is long term. A Marine successfully completing the assignment or extension will be administratively removed from the BCP (and RPCP).

a. Sixth Endorsement (appendix D, enclosure 7). The CO will complete the Sixth Endorsement, counsel the Marine on having successfully completed the BCP assignment, and have the Marine sign the endorsement. The Seventh Endorsement (appendix D, enclosure 8) only applies to Marines completing an extension to a first assignment.

b. The CO will direct the Admin Section to officially remove the Marine from the BCP, recording the following code/statement per reference (e) into MCTFS: TTC 339 000 FR WT CNTL. The Admin Section will record the Unit Diary number and date on the bottom of the endorsement and forward it to the Command PT Representative to be maintained in the Marine's BCP file.

c. Copies of completed BCP files will be kept on record and maintained by the Command PT Representative for 1 year.

14. Summary of Administrative Actions. Table 3-1 lists the administrative actions required for Marines who fail to meet the weight and body fat standards.

ADMINISTRATIVE ACTION	NUMBER OF ASSIGNMENTS	
	BCP 1st	BCP 2nd
Mandatory RPCP	YES	YES
Page 11 SRB/OQR Mandatory Entry	YES*	YES*
Adverse FitRpt or Substandard PRO/CON Marks	YES	YES
Eligible for Promotion	NO	NO
Eligible for PCS Transfer	YES	NO
Process for Administrative Separation	YES	YES
May Be Denied Re-enlistment	YES	YES
May Be Denied Special Schools	YES	YES
May Be Denied Recommendation for Promotion	YES	YES

Table 3-1.--Administrative Actions.

*A 6105 counseling entry will be recorded on the SRB/OQR's Page 11 the day the Marine is counseled on deficiencies, corrective actions to be taken, and assignment to the BCP. Additional Page 11 entries, per paragraph 6206 of MCO P1900.16F will only be required for a Marine who have not made satisfactory progress during the BCP assignment.

NOTE: A Marine who meets the Physical Performance Evaluation criteria will not be assigned to the BCP and will not receive an adverse FitRpt.

15. BCP and End of Active Service (EAS). CO discretion will be exercised when there are individual cases involving Marines who fail to meet weight and body fat standards just prior to their EAS.

a. Imminent EAS does not preclude a Marine who exceeds weight and body fat standards from being evaluated and processed for a BCP and RPCP assignment. A CO may evaluate and assign a Marine to participate in the BCP and RPCP up to the Marine's EAS date.

b. For a Marine who fails to meet weight and body fat standards upon completion of their 6-month BCP assignment or extension, and EAS is imminent, CO discretion will be exercised to decide whether or not measures to process the Marine for administrative separation is worth the administrative burden. CO may opt to extend the Marine's BCP and RPCP assignment up to the Marine's EAS date. A 6105 counseling entry on the SRB/OQR's Page 11 will be recorded, identifying the BCP failure, for Reserve component purposes.

16. Relocation Orders for Marines on the BCP

a. A Marine can be transferred, receive PCS or PCA orders during a first assignment, or (first assignment) extension to a BCP.

b. A Marine will not be transferred, but placed in an administrative hold status if he/she is currently participating in a second assignment to the BCP.

c. The Admin Section and Command PT Representative are required to ensure all Unit Diary entries, MO's Body Composition Evaluation(s), and BCP documents, are up to date and included in the Marine's BCP file, prior to being forwarded to the receiving command.

d. The command is required to forward by registered/receipt mail, the original BCP file to the command to which the Marine is being transferred, and keep a copy of the file on record for 1 year. This BCP file should be forwarded within 7 days of the Marine's official checkout date.

3202. Residential Obesity Program (ROP)

1. Marines that have been evaluated by a MO and determined to have an extreme weight and body fat increase due to an underlying or associated disease, will be recommended to participate in the ROP. The ROP is available to all Marines regardless of pay grade. A Marine will be sent to the ROP at the local ARC based on the following criteria:

a. A Marine has not previously participated in a ROP at a residential ARC during Marine Corps career.

b. A Marine has no medical history of having an eating disorder, (e.g., anorexia, bulimia, or other unresolved eating disorder).

c. A Marine must have a strong potential for continued military service supported by CO or related letters of recommendation.

d. A Marine must have at least 12 months time in service remaining on current military contract after the completion of the ROP for aftercare. [Waivers to remaining time in service criteria may be requested on a case-by-case basis from CMC (MMSR-4).]

2. Post-Residential Obesity Program (Rehabilitation). Upon completion of the ROP, the Marine will be assigned to the unit's RPCP for a minimum of 6 months. Weight and body fat reduction goals will vary depending upon the Marine's circumstances. Results must be monitored by the Command PT Representative and the MO assigned to the case. Entry in the ROP does not affect the BCP assignment process.

3. Rehabilitation Failure. In cases of rehabilitative failure, a Marine will be processed for administrative separation per paragraph 6206.1 of MCO P1900.16F. All rehabilitation failures must be reported to CMC (MMSR-4) via Naval message. Rehabilitation failure is based on the following circumstances:

a. Failure to demonstrate significant and consistent progress towards meeting Marine Corps weight and body fat standards during the 6-month rehabilitation period.

b. Continued increase in weight and/or body fat during the ROP or the aftercare period.

APPENDIX A

GLOSSARY

SECTION 1: MCO SHORT TITLES

MARCORMAN	THE MARINE CORPS MANUAL
MCO P5600.31G	MARINE CORPS PUBLICATIONS AND PRINTING REGULATIONS
MCO P1610.7E	PERFORMANCE EVALUATION SYSTEM (SHORT TITLE: PES)
MCO P1080.40C	MARINE CORPS TOTAL FORCE SYSTEM PERSONNEL REPORTING INSTRUCTIONS MANUAL (SHORT TITLE: MCTFSPRIM)
MCO P1070.12K	MARINE CORPS INDIVIDUAL RECORDS ADMINISTRATION MANUAL (SHORT TITLE: IRAM)
MCO P1900.16F	MARINE CORPS SEPARATION AND RETIREMENT MANUAL (SHORT TITLE: MARCORSEPMAN)
MCO P1080.20M	MARINE CORPS TOTAL FORCE SYSTEM CODES MANUAL (SHORT TITLE: MCTFSCODESMAN)
MCO P1040.31H	ENLISTED CAREER PLANNING AND RETENTION MANUAL
MCO P1400.31B	MARINE CORPS PROMOTION MANUAL, VOLUME 1, OFFICER PROMOTIONS (SHORT TITLE: MARCORPROMMAN, VOL 1, OFFPROM)
MCO P1400.32C	MARINE CORPS PROMOTION MANUAL, VOLUME 2, ENLISTED PROMOTIONS (SHORT TITLE: MARCORPROMMAN, VOL 2, ENLPROM)
MCO 5000.12D	MARINE CORPS POLICY ON PREGNANCY AND PARENTHOOD

SECTION 2: SHORT TITLES & ACRONYMS

APPROPRIATELY CREDENTIALED HEALTH CARE PROVIDER	ACHCP
ADMINISTRATIVE SEPARATION	ADSEP
ALCOHOL REHABILITATION CENTER	ARC
BODY COMPOSITION PROGRAM	BCP
CONVENIENCE OF THE GOVERNMENT	CofG
DISABILITY EVALUATION SYSTEM	DES
DIARY FEEDBACK REPORT	DFR
MARINE CORPS TOTAL FORCE SYSTEM	MCTFS
MEDICAL EVALUATION BOARD	MEB
MEDICAL OFFICER	MO
MILITARY TREATMENT FACILITY	MTF
NOT MEDICALLY QUALIFIED	NMED
PHYSICAL CONDITIONING PROGRAM	PCP
PHYSICAL EVALUATION BOARD	PEB
PERMANENT LIMITED DUTY	PLD
PARTIAL PHYSICAL FITNESS TEST	PPFT
PHYSICAL READINESS TRAINING	PRT
REQUIRED BUT DID NOT TAKE	RDNT
RESIDENTIAL OBESITY PROGRAM	ROP
REMEDIAL PHYSICAL CONDITIONING PROGRAM	RPCP
RETURN TO FULL DUTY	RTFD
TEMPORARY LIMITED DUTY	TLD

SECTION 3: DEFINITIONS

Appropriately Credentialed Health Care Provider (ACHCP). An individual who has specialized training or education in a health-related field. This may include administration, direct provision of patient care, or ancillary services. Such a professional is licensed, certified, or registered by a Government agency to provide specific health services in the field as an independent. An ACHCP, referred to throughout MCO P6100.12 as Medical Officer (MO), is a physician, physician's assistant, or nurse practitioner; Independent Corpsmen or regular Corpsmen are not considered ACHCPs.

Body Composition Program (BCP). Formerly known as the "Weight Control Program." Marines assigned to the BCP will receive assistance in reducing weight/body fat, in order to resume to a more healthy physical fitness state. The BCP assignment process and endorsements are contained in appendix D; the unit's administrative actions are contained in paragraph 3201 of MCO P6100.12.

Body Fat (BF). The amount of fat tissue a person has in ratio to lean body mass (or muscles, bones, organs, nervous tissue, and skin).

Diary Feedback Report (DFR). A report received in the Admin Section requiring action (e.g., DFR to notify the command that SNM's BCP assignment period has expired).

Light Duty. This is the status a MO assigns to a Marine when it is determined that a medical condition exists that interferes with the performance of duty, but the Marine is expected to return to full duty (RTFD). A MO may recommend up to 30 days light duty. When the light duty period expires, SNM will be medically reevaluated. If SNM is not RTFD status, then the Marine will be placed on medical hold and a Medical Evaluation Board (MEB) must be initiated and completed within the following 30 days.

Limited Duty. This is the status a MO assigns to a Marine when it is determined that a medical condition exists and will require an extended period of time before being RTFD. See Temporary Limited Duty and Permanent Limited Duty.

MEB. A MEB is a group of MOs at a local Military Treatment Facility (MTF) convened for the purpose of identifying the physical and mental qualifications of referred Marines to RTFD. The MEB is charged with determining whether a Marine's limitation(s) preclude their RTFD within a reasonable period of time or at all. A MEB will evaluate and report on the diagnosis, prognosis, possible RTFD, plan for further treatment, and make medical recommendation(s) for disposition.

MO. An ACHCP; see ACHCP.

Partial Physical Fitness Test (PPFT). Marines who are medically excused from performing a particular event of the PFT, as certified by a MO, are required to perform a PPFT. Marines will be required to obtain at least the minimum passing performance for the two events they are medically qualified, in order to pass the PPFT.

Permanent Limited Duty (PLD). PLD is a continuation on active duty in a permanent limited duty status, which is approved after a Marine has been determined "Unfit" by a PEB as a result of physical disability. PLD is assigned when a Marine is not expected to RTFD earlier than 16 months or it is projected that the condition cannot be corrected. A Marine can remain on active duty with a PLD status, provided criteria is met per MCO P1900.16F. PLD is requested through the PEB and approved by CMC (MMSR-4). PLD requests will be authorized by the PEB based on the best interest of the Marine Corps and the health of the Marine.

Physical Evaluation Board (PEB). The PEB acts on behalf of the SecNav to make determinations of fitness for continued military service, entitlement to benefits, disability ratings, and disposition of the medical case. A PEB is required to convene for cases that exceed the maximum 16 months of limited duty or when a Marine's initial diagnosis and prognosis do not anticipate a return to RFTD.

Physical Fitness Test (PFT). An instrument that measures the level of physical fitness of all Marines. It is a measurement of general fitness vice combat readiness and MOS capability. The standard PFT consists of three events that measure cardiovascular endurance, muscular strength and endurance, and mobility. Male Marines will perform dead-hang pullups, abdominal crunches, and a 3.0 mile run. Female Marines will complete the flexed arm hang, abdominal crunches, and a 3.0 mile run. These events are designed to test the strength and stamina of the upper body (shoulder girth), midsection, and lower body, as well as the efficiency of the cardiovascular system.

Physical Readiness Training (PRT). This training is comprised of general physical conditioning exercises that will develop and maintain strength, endurance, and the physical skills necessary to sustain individual Marines during combat.

Remedial Physical Conditioning Program (RPCP). Remedial physical conditioning is a process by which Marines are conditioned in a progressive manner to meet prescribed standards. The Program affords Marines the opportunity to rebuild themselves after a weight/body fat increase, injury/illness, pregnancy, or a period that lacked a structured fitness program.

Residential Obesity Program (ROP). Marines that have been evaluated by a MO and determined to have an extreme weight/body fat increase due to an underlying cause or associated disease, will be recommended to participate in the ROP.

Temporary Limited Duty (TLD). TLD is an assignment of limited duty for a specific period. Assignment to TLD can only be made after the completion of a MEB, and only if the prognosis is that the Marine can be RTFD within the specified period. The cumulative period of TLD will not exceed 16 months.

APPENDIX B

DAILY 16 PROGRAM

A. General

1. Purpose. The Daily 16 Program is a comprehensive series of warmup, conditioning, and cool-down exercises; it replaces the former Daily 7. The Daily 16 warmup exercises facilitate gradual distribution of blood flow to the muscles, preparing both the cardiovascular and musculoskeletal systems for the exercise session, by effectively targeting both the upper and lower body. The increased blood flow to the muscles produces a warming effect, increasing the elasticity of the muscles and connective tissues, which is believed to reduce injury risks. The Daily 16 conditioning exercises include traditional calisthenics that are safe in providing a total body workout, which can vary in duration, degree of difficulty, and level of intensity. The Daily 16 cool-down exercises (the same exercises used in the warmup) allow the body to gradually return to the pre-exercise state.

2. Scope. There are three different ways the Daily 16 Program can be used during physical training:

a. Daily 16 Warmup (D16WU). A series of warmup and dynamic stretching exercises that should be conducted prior to the main activity (e.g., formation run, obstacle course, circuit course, etc.), of every physical training session.

b. Daily 16 Workout (D16WO). A series of conditioning exercises that can be used as augmentation to another conditioning activity (circuit course, Physical Fitness Test (PFT), etc.), or can be used as a conditioning activity in of itself by simply increasing the number of repetitions or by slowing down the execution of the repetition.

c. Daily 16 Cool-Down (D16CD). A series of cool-down exercises that should be used as the final activity of a physical training session.

3. Daily 16 Warmup and Dynamic Stretching Descriptions

a. Warmup Exercises

(1) Heel to Toe Rocking. Starting position is standing with feet together and hands on hips. Rock back onto the heels, pause, then rock forward onto the toes and pause. Repeat 10 to 15 repetitions.

1

2

(2) Partial Squats. Starting position is standing with feet shoulder width apart, arms at the sides (1). Keeping the heels on the deck, partially squat until hands are near mid-calf. The knees should only bend to about 60 degrees, just short of a sitting position (2). Return to the starting position (3). Repeat 10 to 15 repetitions.

1/3

2

(3) Butt Kicks. Starting position is standing with feet shoulder width apart, hands on hips. Shift weight onto the right foot and quickly bend the left knee five times, bringing the left heel toward the buttocks. Switch legs and repeat. Repeat the whole cycle two or three times, until a total of 15 to 20 repetitions are done on each leg.

(4) Double Time in Place. Starting position is standing with arms at the sides. Slowly begin to run in place and gradually increase speed. While double-timing, conduct "punches to the front (1)" (throw easy punches to the front of the body). Switch to "punches to the sky (2)" (throw easy punches straight up to the sky). Finish with "arm circles (3)", (begin small then transition to large arm circles; repeat in other direction).

1

2

3

(5) Neck Flexion and Extension. Starting position is standing with feet shoulder width apart, hands on hips. Flex the neck forward bringing the chin toward the chest, pause. Extend the head back and pause. Repeat for 5 to 10 repetitions.

(6) Neck Lateral Flexion. Starting position is standing with feet shoulder width apart, hands on hips. Tilt the head to the left side, bringing the left ear toward the left shoulder, pause. Switch sides and repeat. Repeat for 5 to 10 repetitions.

(7) Trunk Flexion and Extension. Starting position is standing with feet shoulder width apart, hands on hips (1). Extend the trunk backward bringing a slight hyperextension of the trunk and pause (2). Repeat for 5 to 10 repetitions.

1

2

(8) Trunk Lateral Flexion. Starting position is standing with feet shoulder width apart, hands on hips. Bend the trunk to the left side (1), pause, then switch to the right side (2) and pause. Repeat for 5 to 10 repetitions. Proceed to the Stretching Exercises.

1

2

b. Dynamic Stretching Exercises

(1) Upper Back Stretch. Starting position is standing with feet shoulder width apart. Extend the arms and clasp the hands in front of the chest. Push the arms forward rounding the shoulders and upper back. Hold the position for 10 seconds and breath naturally. The stretch should be felt over the upper back.

(2) Chest Stretch. Starting position is standing with feet shoulder width apart. Clasp hands together behind the lower back, palms up. Pull the arms up toward the head. Hold the position for 10 seconds and breath naturally. The stretch should be felt in the front of the chest and shoulders.

(3) Modified Hurdler Stretch. Starting position is in the sitting position. Extend the left leg out while tucking the right leg in front of the hips with the knee pointing outward. Bend the torso forward toward the left knee. The stretching should be felt over the back of the left thigh. Hold the position for 10 seconds and breath naturally. Switch sides and repeat.

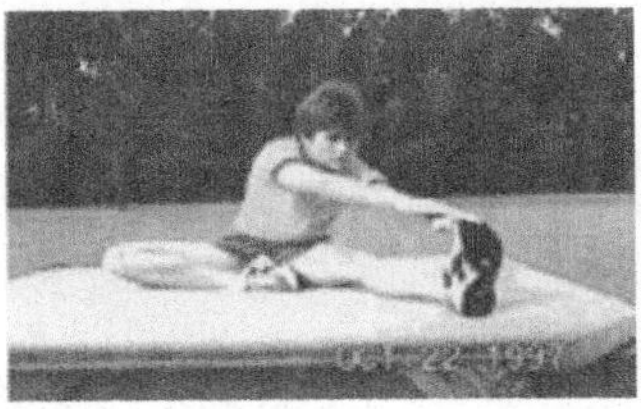

(4) Hip and Back Stretch. Starting position is in the sitting position. Extend the right leg straight out and cross the left leg over the right leg by bending the left knee and placing the left foot on the deck next to the right knee. Turn the upper torso to the left pushing the left knee to the right with the right elbow. Hold the position for 10 seconds and breath naturally. The stretch should be felt over the lower back and left hip. Switch sides and repeat.

(5) Groin Stretch. Starting position is in the sitting position with both knees bent and the bottoms of the feet together. Grasp the feet and gently push the knees with the elbows toward the deck. Hold the position for 10 seconds and breath naturally. The stretch should be felt over the inside of both thighs.

(6) Calf Stretch. Starting position is standing with arms at the sides. Place the left foot approximately 2 feet forward and slightly bend the right knee. Lean forward toward the left foot pointing the left toes up to the sky. Hold the position for 10 seconds and breath naturally. Grabbing the left foot and gently pulling it towards you can increase the level of intensity. The stretch should be felt over the left calf. Switch sides and repeat.

(7) Iliotibial Band (ITB) Stretch. Starting position is standing with arms at the sides. Place the left foot behind and a few inches to the right of the right foot. Bring the left arm over the head. Place your body weight on the left leg and bend at the waist to the right. Hold the position for 10 seconds and breath naturally. The stretch should be felt over the left hip. Switch sides and repeat.

(8) Hamstring Stretch. Starting position is lying down with the back flat against the deck. Bring the left knee toward the chest grasping the left leg just below the knee. Gently straighten the left knee and hold for the count. The right leg should remain on the deck. Hold the position for 10 seconds and breath naturally. The stretch should be felt on the back of the left thigh. Switch sides and repeat.

(9) Neck Stretch. Starting position is standing with feet shoulder width apart. Move both arms behind the back and grasp the left wrist with the right hand. Tilt the head to the right and pull the left arm to the right. Hold the position for 10 seconds and breath naturally. The stretch should be felt over the left shoulder and left side of the neck. Switch sides and repeat.

(10) Triceps Stretch. Starting position is standing, arms at the sides. Bend the left elbow and bring the left arm up and back placing the left hand between the shoulder blades. Gently pull the left elbow with the right hand behind the head. Hold the position for 10 seconds and breath naturally. The stretch should be felt over the back of the upper arm. Switch sides and repeat.

(11) Quadriceps Stretch. Starting position is lying down on the left side. Bend the left hip and knee to 90 degrees. Grasp the right ankle with the right hand and pull the right knee straight back. Do not hyperextend the lower back. Hold the position for 10 seconds and breath naturally. The stretch should be felt over the front of the right thigh. Switch sides and repeat.

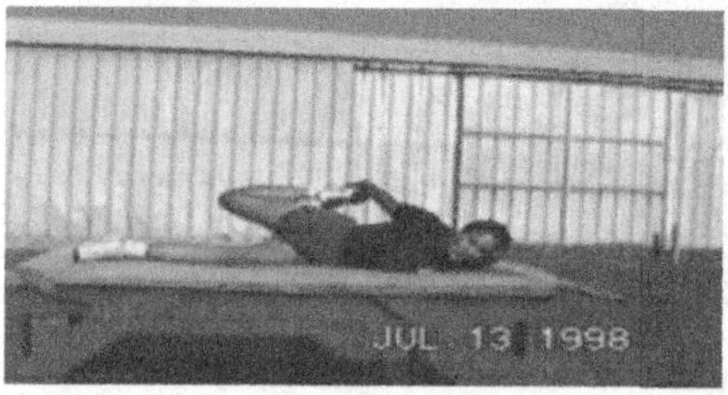

(12) Lying Down ITB Stretch. Starting position is lying down on the deck. Bring the left leg with the knee straight across the body (1). Hold the position for 10 seconds and breath naturally. The stretch should be felt over the left hip. Switch sides and repeat (2).

1

2

(13) Posterior Shoulder Stretch. Starting position is standing with feet shoulder width apart, arms at the sides. Bend the left elbow and bring the left arm across the chest. Give a gentle pull with the right hand. Hold the position for 10 seconds and breath naturally. The stretch should be felt over the posterior left shoulder. Switch sides and repeat.

(14) Hip Flexor Stretch. Starting position is standing, hands on hips. Step the left foot forward 3 to 4 feet. Place the right knee on the deck. Gently move the left knee forward. Hold the position for 10 seconds and breath naturally. The stretch should be felt over the front of the right thigh and hip. Switch sides and repeat.

(15) Single-Leg Lower Back Stretch. Starting position is lying with the back flat against the deck. Bring the right knee toward the chest grasping the right knee. Gently pull the knee tight into the chest. The left leg should remain on the deck. Hold the position for 10 seconds and breath naturally. The stretch should be felt along the lower back to the right buttock. Switch sides and repeat.

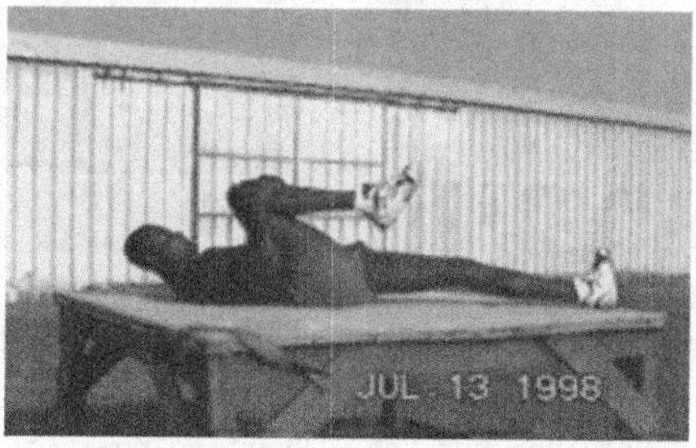

(16) Prone Abdominal Stretch. Starting position is lying on the stomach with the hands placed near the shoulders as if in the down position of a pushup. Slowly raise the upper body, keeping the waist on the deck. Hold the position for 10 seconds and breath naturally. The stretch should be felt over the abdomen.

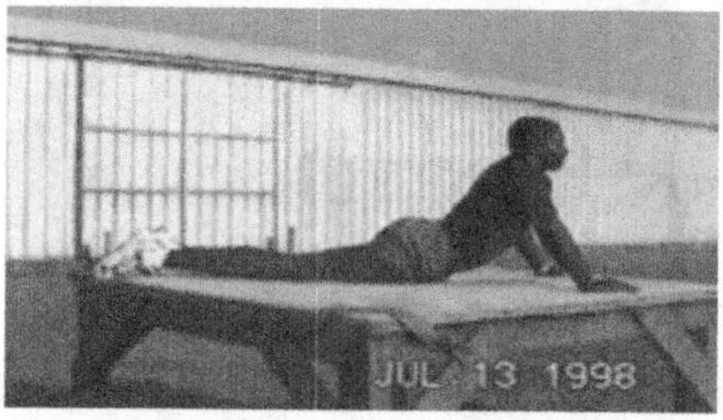

4. Daily 16 Conditioning Exercise Descriptions

a. Pushups. Starting position is lying on your stomach with hands shoulder width apart, toes on the deck, and elbows, back, and knees straight. On the first and third count, lower the chest to the deck; bend the elbows to at least 90 degrees (1/3). On the second and fourth count, extend the arms back to the starting position (2/4). This exercise conditions the chest, primarily in the anterior shoulder region, and secondarily, the triceps.

1/3

2/4

b. Crunches. Starting position is lying on the back with the hips bent to 90 degrees and the knees bent, feet off the deck. Bend the elbows to 90 degrees and fold across the chest or rib cage. On the first and third count raise the upper torso off the deck touching the thighs with the forearms (1/3). On the second and fourth count, return to the starting position (2/4). The arms must remain in constant contact with the chest/rib cage when executing the crunch. This should be done in a slow and controlled manner. This exercise conditions the abdominal muscles.

1/3

2/4

c. Dirty Dogs. Starting position is on the hands and knees. On the first and third count raise the left leg to the side, while keeping the knee bent (1/3). On the second and fourth count return the leg to the starting position (2/4). Switch sides and repeat. This exercise conditions the hip abductors.

1/3

2/4

d. Wide Pushups. Starting position is lying on the stomach with hands beyond shoulder width apart, toes on the ground, and elbows, back, and knees straight. On the first and third count, lower the chest to the deck, bending the elbows to at least 90 degrees (1/3). On the second and fourth count, extend the arms back to the starting position (2/4). This exercise conditions the chest, primarily in the anterior shoulder region, and secondarily, the triceps. With the wider hand position, the chest muscles increase their workload.

1/3

2/4

e. Dive Bomber Pushups. Starting position is lying on the stomach with hands and toes on the deck, and elbows and knees straight. The hands will be slightly beyond shoulder width apart. The hips will be raised up and the shoulders will be behind the hands. On the first count, lower the chest down and forward to the deck, the shoulders will be even with the hands (1). On the second count, continue forward extending the elbows where now the shoulders are in front of the hands (2). On the third count, reverse the direction lowering the chest down and back to the deck, the shoulders will be even with hands (3). On the fourth count, continue back and up to the starting position (4). This exercise is done in a smooth, continuous motion.

This exercise conditions the chest and anterior shoulder primarily, and secondarily, the triceps through a greater ranger of motion.

1

2

3

4

f. Elbow to Knee Crunches. Starting position is lying on the back with the right foot flat on the deck and, the left foot crossed over the right knee, and the arms crossed over the chest. On the first and third count, raise the upper torso off the deck rotating to the left touching the right elbow to the left thigh (1/3). On the second and fourth count, return to the starting position (2/4). This exercise should be done in a slow and controlled manner. Switch sides and repeat. This exercise conditions the abdominal muscles with more emphasis on the oblique.

1/3

2/4

g. Side Crunches. Starting position is lying on the left side with the left arm across the chest and right arm along the side of the body. On the first and third count, raise the upper torso and feet off the deck sliding the right hand down the thigh (1/3). On the second and fourth count, return to the starting position (2/4). Switch sides and repeat. This exercise conditions the abdominal muscles with emphasis on the internal oblique and external oblique.

1/3

2/4

h. Prone Flutter Kicks. Starting position is lying on the stomach. On the first count raise the left leg off the deck while the right leg remains on the deck (1). On the second count, lower the left leg to the starting position (2). On the third count, raise the right leg off the deck while the left leg remains on the deck (3). On the fourth count, lower the right leg to the starting position (4). This exercise conditions the muscles that extend the hip and back.

1

2

3

4

i. Back Extension. Starting position is lying on the stomach with hands behind the head. On the first and third count, raise the upper torso and legs off the deck (1/3). On the second and fourth count, lower the upper torso and legs to the starting position (2/4). This exercise conditions the muscles that extend the back.

1/3

2/4

j. Donkey Kicks. Starting position is on the hands and knees. On the first and third count, kick the left leg back and up straightening the knee (1/3). On the second and fourth count, bend the knee and hip bringing the left knee into the chest (2/4). The back should not hyperextend during this exercise. Switch sides and repeat. This exercise conditions the muscles that extend the hip.

1/3

2/4

k. Hip Adduction. Staring position is lying on the left side with the right leg bent setting the right foot in front of the left knee. On the first and third count, raise the straight left leg off the deck squeezing the thighs together (1/3). On the second and fourth count, lower the left leg to the starting position (2/4). The left toes should be pointing straight-forward not to the sky. Switch side and repeat. This exercise conditions the muscles on the inner thigh.

1/3

2/4

1. <u>Side Leg Raises</u>. Starting position is lying on the left side with the left knee bent, the hip vertical and the toes on the right foot pointing forward not to the sky. On the first and third count, raise the right leg approximately 18 inches leading with the heel (1/3). The toes will still point forward, not to the sky. On the second and fourth count, lower the right leg to the starting position (2/4). Switch sides and repeat. This exercise conditions the muscles on the side of the hip and thigh.

1/3

2/4

m. Steam Engines. Starting position is standing with the feet shoulder width apart and hands behind the head. On the first count, touch the right elbow to the left knee by bending and raising the left knee and twisting and bending the upper torso to the left (1). On the second count, return to the starting position (2). On the third count, touch the left elbow to the right knee by bending and raising the right knee and twisting and bending the upper torso to the right (3). On the fourth count, return to the starting position (4). This exercise should be done in a slow and controlled cadence. This exercise conditions the abdominal muscles, primarily the oblique, and secondarily, the hip and thigh.

1

2

3

4

n. Lunges. Staring position is standing with the feet shoulder width apart and hands on the hips. On the first count, touch the right knee to the deck by stepping forward with the left foot and bending both knees (1). On the second count, return to the starting position (2). On the third count, touch the left knee to the deck by stepping forward with the right foot and bending both knees (3). On the fourth count, return to the starting position (4). Do not bend the forward knee more than 90 degrees. This exercise should be done in a slow and controlled cadence. This exercise conditions the muscles that extend the hip and knee of the forward leg.

1

2

3

4

o. Side Straddle Hops. Starting position is standing with the feet together and arms at the sides. On the first and third count, jump and land with both feet just beyond shoulder width apart while bringing the hands together overhead (1/3). On the second and fourth count, jump back to the starting position (2/4). This exercise is done in a smooth, continuous motion. This exercise conditions the upper arm, hip, and thigh.

1/3

2/4

5. Daily 16 Warmup/Cool-Down, Stretching and Exercise Cards

STRETCHING CARD A

Chest Stretch
Triceps Stretch
Posterior Shoulder Stretch
Iliotibial Band (ITB) Stretch
Modified Hurdler Stretch
Hip and Back Stretch
Quadriceps Stretch
Lower Back Stretch
Abdominal Stretch

STRETCHING CARD B

Upper Back Stretch
Chest Stretch
Shoulder & Neck Stretch
Triceps Stretch
Posterior Shoulder Stretch
Lying Down (ITB) Stretch
Modified Hurdler Stretch
Groin Stretch
Calf Stretch

STRETCHING CARD C

Active Hamstring Stretch
Groin Stretch
Hip Flexor Stretch
Lower Back Stretch
ITB Stretch
Calf Stretch
Neck Stretch
Upper Back Stretch
Chest Stretch

EXERCISE CARD 1

Wide Pushups
Donkey Kicks
Crunches
Dive Bomber Pushups
Dirty Dogs
Side Crunches
Back Extensions
Lunges
Side Straddle Hops

EXERCISE CARD 2

Pushups
Crunches
Side Leg Raises
Diamond Pushups
Elbow-to-Knee Crunches
Prone Flutter Kicks
Hip Abduction
Lunges
Steam Engines

WARM-UP/COOL-DOWN CARD

Toe-Heel Rocking
Partial Squats
Butt Kicks
Trunk Bends
Neck Bends
Run in Place
Punch to the Front
Punch to the Sky
Arm Circles
*Stretch Card (10 Seconds)
*Exercise Cards

APPENDIX C

REMEDIAL PHYSICAL CONDITIONING PROGRAM (RPCP)

1. Purpose. Remedial physical conditioning is a process by which Marines are conditioned in a progressive manner to meet prescribed physical fitness and body composition standards. The RPCP goal is to provide challenging conditioning sessions, using the spectrum of aerobic conditioning, resistance conditioning, and other related exercises that will afford the Marines an opportunity to rebuild themselves after a weight or body fat increase, injury/illness, pregnancy, or a period that lacked a structured fitness program.

2. Assignment to RPCP. Participation in the RPCP is a mandatory requirement for Marines who fail the Physical Fitness Test (PFT) or who are assigned to the Body Composition Program (BCP). Marines will also be assigned to the RPCP when they consistently display problems in MOS-specific tasks, (e.g., conditioning marches, etc.) or struggle to meet the minimum standard for an event during a semi-annual PFT. Performance on unit formation runs will not be used as RPCP entry/exit criteria. Marines recovering from injuries or prolonged illness may be assigned to the RPCP if a formal remedial program would be beneficial. In the later case, COs must take into account any followup medical care (therapy) requirement and tailor the RPCP to accommodate those physical limitations or restrictions imposed by the Medical Officer (MO). Deficiencies, which can be improved or corrected by a command-directed remedial conditioning program, fall into several categories:

a. Lack of strength in one or more body parts;

b. Lack of overall cardiovascular or muscular endurance;

c. Lack of adherence to body composition standards;

d. Lack of mobility (posture, balance, stability, agility, coordination, power, speed and flexibility);

e. Lack of motivation and dedication; and/or

f. Lack of education and experience in health and fitness.

3. Organization of the RPCP

a. Leadership. The intent of the RPCP is not punitive in nature, but designed to encourage improvement in overall fitness and body composition. Unit leaders and Command Physical Training (PT) Representatives should assist Marines in clearly identifying their deficiencies, developing specific strategies to improve, and tracking progress. The RPCP should focus on realistic goal setting, healthy life-styling strategies, progressive PT, dietary measures, and testing and evaluation. The RPCP should be challenging, employ a cross-training philosophy, and ensure safety measures are taken to avoid over-training and potential overuse injuries.

b. Use of Resources. This Appendix will provide a few examples of remedial conditioning programs units can employ or use as a guide for their own program designs. Command PT Representatives can also access the resources available at Semper Fit Fitness Centers, which are staffed by qualified professionals and offer a variety of health and fitness services.

c. Organization. Remedial conditioning should be conducted in small groups, (e.g., no larger than 10-15 Marines per group), thereby allowing for more focused instruction, oversight, and reinforcement. It is recommended that the RPCP initially be organized by grouping Marines by area of deficiency and level thereof. Prescribed exercise loads should be commensurate with their abilities, and whenever possible, linking a strong Marine with a weaker Marine. An organization as such will support performance and improvement.

(1) As the RPCP and individual fitness and performance levels progress, Marines should be reorganized to concentrate on remaining deficiencies, or combined to conduct specific PT events, such as conditioning marches or team sports. For example, a group that exhibits a deficiency in upper body strength would execute a program that includes pullups, rope climbing, rifle PT, log drills, and weight training. A second group may exhibit deficiencies in cardiovascular respiratory endurance, and therefore would execute a program that includes individual and ability group running, grass drills, circuit training, and fartlek running.

d. Measurement. Continual measurement of a Marine's deficient area(s) is the only way to gauge progress. While testing should not take the place of training sessions geared towards improving performance, periodic testing (e.g., every 2 weeks) is an important step of any remedial conditioning program. Testing, PFT, and body composition evaluations for example, should commence at the end of the first week of a RPCP.

4. RPCP

a. Circuit Training and Exercise Drills. Circuit training is described as a specific training routine in which a variety of fitness discrepancies are addressed It can accommodate a large number of Marines with little supervision, provide variety and challenge with its progressive programming, and allow participants to progress at their own rate. The goal of circuit training is to develop strength and endurance through a systematic and progressive conditioning program that involves stations where specific exercises are performed. These exercises are performed vigorously for a short period of time before moving to a follow-on station. The vigorous activity in short periods of time provide a near maximal quality training session, assuring progressive overload in an organized manner. The exercise selected for each station and the arrangement of the stations is determined by the objective of the circuit.

(1) Free Circuit. Free (portable) circuits use equipment that can be readily transported or set up for use regardless of the training area and environment. A simplified course can be executed on a basketball court, a field, or on the deck of a ship. In a free circuit, there is no set time for staying at each station and no signal to move from one station to the next. Marines work at their own pace, doing a fixed number of repetitions (rep) at each station. Progress is measured by the time needed to complete a circuit. However, in remedial training, the quality and number of repetitions should be monitored.

(2) Fixed Circuit. Fixed circuits require equipment that is normally in place and can be used on a regular basis. In a fixed circuit, a specific length of time is set for each station. The time is monitored with a stopwatch, with Marines rotating through the stations on command. To increase the intensity or difficulty of a fixed circuit:

(a) maintain the time for completion, but increase the number of reps;

(b) increase the time per station along with the number of reps; and,

(c) increase the number of times a circuit is negotiated.

b. Number of Stations/Circuit Training Reps. The goal of the circuit, as well as the time and equipment available will determine the number of stations. A circuit geared for the objective of developing upper or lower body strength may need as little as six to eight stations, while a cardiovascular strength and endurance circuit may have as many as 20 stations. The number of Marines assigned to negotiate a single circuit station is determined by the total number of participants, the number of stations, and the level of supervision available. To achieve the desired training effect, the circuit may have to be repeated several times. For example, a 10-station circuit requiring Marines to exercise for 30 seconds at each station and 15 seconds to move between stations may result in Marines completing the entire evolution three times. The entire training session, designed to take between 45-60 minutes, includes a warmup and cool-down period (e.g., Daily 16). As the conditioning level increases, consideration can be given to either increase exercise periods from 30 to 45 and 60 second intervals, or the addition of a fourth rotation through the circuit. Stations should be arranged in a sequence that allows for adequate recovery time between high/low intensity exercises or by alternating different muscles groups.

(1) Cardiovascular Endurance Circuit #1. This circuit is designed to improve the cardio-respiratory system by exercising hard at each of the eight exercise stations, with a steady recovery time (slow jog) between stations. Recommended course length is approximately 1/4 to 1/2 mile in length.

Station #1:	Ab Crunch	20 reps
Station #2:	Bend and Thrust	20 reps
Station #3:	Sprint	1 rep
Station #4:	Inclined Pushups	20 reps
Station #5:	Star Jumps	20 reps
Station #6:	Dorsal Raise	20 reps
Station #7:	Knee Lifts	20 reps
Station #8:	Dips	20 reps

(2) Cardiovascular Endurance Circuit #2. Recommended course length is 1/2 to 1.0-mile, with 14 events/stations:

Station #1:	Stationary Run	30 seconds
Station #2:	Pushup	30 seconds
Station #3:	Side Straddle Hop	30 seconds
Station #4:	Crunches	30 seconds
Station #5:	Ski Jumps	30 seconds
Station #6:	Flutter Kicks	30 seconds
Station #7:	Bend and Reach (slow reps)	30 seconds
Station #8:	Wide Pushups	30 seconds
Station #9:	Bicycle	30 seconds
Station #10:	Knee Bender	30 seconds
Station #11:	Steam Engines (Daily 16)	30 seconds
Station #12:	Twisting Situp	30 seconds
Station #13:	Lunges (Daily 16)	30 seconds
Station #14:	All-Fours Run	30 seconds

(3) Upper Body Strength Circuit #2. This circuit is designed to improve upper body strength and endurance, and is normally conducted following a cardiovascular training session. Proper form and full range of motion are important. Eight stations are set up over a course length of approximately 500 yards.

Station #1:	Crunches	20 reps
Station #2:	Pullups	3 sets @ max effort
Station #3:	Dorsal Raises	20 reps
Station #4:	Tricep Dips	40 reps
Station #5:	Situps	2 sets of 5 reps
Station #6:	Rope Heaves	20 reps
Station #7:	Pushups	20 reps
Station #8:	Sprint	1 rep

(4) Upper Body/Abdominal Strength Circuit #3. This circuit is designed to encompass a cardiovascular/anaerobic (sprint) portion that incorporates eight exercise stations with no set course length required. This circuit is conducted without a set time signal, e.g., Marines move immediately to a station after each exercise is completed. Recommended one to two complete course rotations, with a possible time decrease to 20 seconds (vice 30) on the second running.

Station #1:	Elevated Pushup	30 seconds
Station #2:	Twisting Situp	30 seconds
Station #3:	Parallel Dips	30 seconds
Station #4:	Ab Crunch (feet unsecured)	30 seconds
Station #5:	Wide Pushup	30 seconds
Station #6:	Flutter Kicks	30 seconds
Station #7:	Diamond Pushup	30 seconds
Station #8:	Bicycle	30 seconds

(5) Lower Body/Upper Body Strength (Weight Room) Circuit #4. This circuit is designed to follow a cardiovascular session. There are 13 exercise stations and the circuit should be conducted in one to two complete rotations. Each exercise should encompass slow and controlled movements, striving for muscle failure within 8-12 repetitions. A preliminary session, for Marines to become familiar with weight room layout and starting weight limits and execution is recommended.

Station #1:	Leg Press	8-12 reps
Station #2:	Leg Raise	8-12 reps
Station #3:	Leg Extension	8-12 reps
Station #4:	Leg Curl	8-12 reps
Station #5:	Heel Raise	8-12 reps
Station #6:	Bench Press	8-12 reps
Station #7:	Seated Row	8-12 reps
Station #8:	Military Press	8-12 reps
Station #9:	Lat Pulldown	8-12 reps
Station #10:	Shrugs	8-12 reps
Station #11:	Triceps Extension	8-12 reps
Station #12:	Biceps Curl	8-12 reps
Station #13:	Incline Situp	8-12 reps

c. Fartlek Training. Fartlek training is designed to develop strength, stamina, and endurance by combining running (cardiovascular) and calisthenics (strength and endurance). Fartleks consist of various running exercises, which strive to overload the cardiovascular system by allowing only partial recovery between each intense training segments; this sequence is repeated for the duration of the training session. Fartleks are normally conducted over a minimum 3-mile course, preferably with varying terrain and intermittent exercise stations to create the work overload.

(1) Initial/Baseline Fartlek:

- Warmup (Daily 16 warmup routine)
- Warmup run at easy pace for 5-10 minutes
- Run fast, steady pace for 3/4-1 3/4 miles
- Walk/Jog at moderate pace for 5 minutes (recovery)
- Run easily - sprinting 15-20 meters occasionally
- Run full speed uphill for 175-200 meters
- Cool-down by running easily for 1/2-1 mile
- Cool-down (Daily 16 cool-down routine)

(2) Cardiovascular/Muscular Endurance Fartlek: Recommended course length is 3.0-3.5 miles. Location and spacing of exercise stations is normally based on terrain. After a unit orientation run, the Fartlek course should be run as an individual effort.

Station #1:	Mountain Climbers	20 reps
Station #2:	Bends and Thrust	20 reps
Station #3:	Ab Crunches (feet unsecured)	20 reps
Station #4:	Sprint	175-200 meters
Station #5:	Situp w/Twist	20 reps
Station #6:	Pullups or Pushups	10-20 reps
Station #7:	Dorsal Raises	20 reps
Station #8:	Star Jumps	20 reps
Station #9:	Squat Thrusts	20 reps

5. RPCP Schedules

a. Upper Body Strength (Weekly) Circuit. A 3 day per week routine is recommended to improve overall upper body strength deficiencies. Conduct tests every 2 weeks and increase weights and reps as necessary.

Monday: Weight Training

Pullups	2 sets to fatigue
Lat Pulldown*	2 sets of 15 reps
Seated Row*	2 sets of 15 reps
Biceps Curl (EZ-Curl Bar)*	2 sets of 15 reps
Alternating Bicep Curls (Dumbbell)	2 sets of 15 reps

* Strive for momentary muscular fatigue at the end of each set. Allow 45-60 seconds rest between sets and 2-3 minutes rest between exercises.

Tuesday: Cross Training - Cardiovascular/Abdominal Training

Wednesday: Pullup/Pushup/Crunch Pyramid **

Set #1:	1 Pullups/2 Pushups
Set #2:	2 Pullups/4 Pushups
Set #3:	3 Pullups/6 Pushups/25 Crunches
Set #4:	4 Pullups/8 Pushups
Set #5:	5 Pullups/10 Pushups/25 Crunches
Set #6:	4 Pullups/8 Pushups
Set #7:	3 Pullups/6 Pushups/25 Crunches
Set #8:	2 Pullups/4 Pushups
Set #9:	1 Pullups/2 Pushups

** Decrease/Increase reps in pyramid based on the Marine's ability and progression of strength, the minimal rest between sets, and spotter assistance necessary to complete required reps.

Thursday: Rest

Friday: Repeat Monday training session

b. Strength and Endurance Training Schedule (20 weeks). This training schedule's primary fitness element is running, but also strives to combine aspects of strength, endurance, and flexibility. The running portion of the program provides the base of cardiovascular endurance necessary to augment all other physical skills. In addition to the progressive overload running program, strength training and flexibility exercises (e.g., Daily 16), addressing lower and upper body resistance training must be incorporated. Weight training can augment or be substituted for the Daily 16 exercises. Each training period will include a warmup prior to exercise and a cool-down session to follow.

WEEK 1,2	Endurance	Strength
Monday	Run 40 min @ 10 min/mile pace	Lower Body Resistance Daily 16 or Weight Training (1 set)
Tuesday	None	Upper Body Resistance - Daily 16 or Weight Training (1 set)
Wednesday	Run 30 min @ 10 min/mile pace	Lower Body - (1 set)
Thursday	None	Upper Body - (1 set)
Friday	Run 30 min @ 10 min/mile pace	Lower Body - (1 set)

WEEK 3,4	Endurance	Strength
Monday	Run 44 min @ 10 min/mile	Lower Body - (1 set)
Tuesday	None	Upper Body - (1 set)
Wednesday	Run 33 min @ 10 min/mile	Lower Body - (1 set)
Thursday	None	Upper Body - (1 set)
Friday	Run 33 min @ 10 min/mile	Lower Body - (1 set)

WEEK 5,6	Endurance	Strength
Monday	Run 48 min @ 10 min/mile	Lower Body - (2 sets)
Tuesday	None	Upper Body - (2 sets)
Wednesday	Run 36 min @ 10 min/mile	Lower Body - (2 sets)
Thursday	None	Upper Body - (2 sets)
Friday	Run 36 min @ 10 min/mile	Lower Body - (2 sets)

WEEK 7,8	Endurance	Strength
Monday	Run 53 min @ 10 min/mile	Lower Body - (2 sets)
Tuesday	None	Upper Body - (2 sets)
Wednesday	Run 40 min @ 10 min/mile	Lower Body - (2 sets)
Thursday	None	Upper Body - (2 sets)
Friday	Run 40 min @ 10 min/mile	Lower Body - (2 sets)

WEEK 9	Endurance	Strength
Monday	Run 53 min @ 9 min/mile	Upper Body - (3 sets)
Tuesday	None	Lower Body - (3 sets)
Wednesday	Run 40 min @ 10 min/mile	Upper Body - (2 sets)
Thursday	None	Lower Body - (3 sets)
Friday	Run 40 min @ 10 min/mile	Lower Body - (2 sets) Power - 2 reps

WEEK 10	Endurance	Strength
Monday	Run 53 min @ 8.5 min/mile	Upper Body - (2 sets)
Tuesday	Intervals: 4x 400m sprints in 2 min	Lower Body - (3 sets)
Wednesday	Run 40 min @ 8.5 min/mile	Upper Body - (2 sets)
Thursday	None	Lower Body - (3 sets) Power - 3 reps
Friday	Run 40 min @ 8.5 min/mile	Upper Body - (2 sets)

WEEK 11	Endurance	Strength
Monday	Run 53 min @ 8.5 min/mile	Upper Body - (3 sets)
Tuesday	Intervals: 4x 400 m sprints in 2 min	Lower Body - (2 sets)
Wednesday	Run 40 min @ 8.5 min/mile	Upper Body - (3 sets)
Thursday	None	Lower Body - (2 sets) Power - 4 reps
Friday	Run 40 min @ 8.5 min/mile	Upper Body - (3 sets)

WEEK 12	Endurance	Strength
Monday	Run 58 min @ 8.5 min/mile	None
Tuesday	Intervals: 5x 400 m sprints in 2 min (4 min rest)	Upper/Lower Body - (2 Sets)
Wednesday	Run 40 min @ 8.5 min/mile	None
Thursday	None	Upper/Lower Body - (3 sets) Power - 4 reps
Friday	Run 30 min @ 7.5 min/mile	None

WEEK 13	Endurance	Strength
Monday	Run 58 min @ 8.5 min/mile	None
Tuesday	Intervals: 6x 400 m sprints in 2 min (4 min rest)	Upper/Lower Body - (3 sets)
Wednesday	Run 40 min @ 8.0 min/mile	None
Thursday	None	Upper/Lower Body - (3 sets) Power - 4 reps
Friday	Run 30 min @ 7.0 min/mile	None

WEEK 14	Endurance	Strength
Monday	Run 58 min @ 8.5 min/mile	None
Tuesday	Intervals: 6x 400 m sprints in 2 min (3:30 min rest)	Upper/Lower Body - (3 sets)
Wednesday	Run 40 min @ 8.0 min/mile	None
Thursday	None	Upper/Lower Body - (3 sets) Power - 4 reps
Friday	Run 30 min @ 7.0 min/mile	None

WEEK 15	Endurance	Strength
Monday	Run 30 min (easy pace)	Upper/Lower - (1 set)
Tuesday	None	None
Wednesday	Run 20 min (easy pace)	None
Thursday	None	None
Friday	Run 20 min (easy pace)	Upper/Lower - (1 set)

WEEK 16	Endurance	Strength
Monday	Run 30 min (easy pace)	Upper/Lower - (1 set)
Tuesday	None	None
Wednesday	Run 20 min (easy pace)	None
Thursday	None	None
Friday	Run 20 min (easy pace)	Upper/Lower - (1 set)

WEEK 17	Endurance	Strength
Monday	Run 30 min (easy pace)	None
Tuesday	None	Lower Body - (2 sets) Power - 2 sets
Wednesday	Run 20 min (easy pace)	None
Thursday	None	Lower Body - (2 sets) Power - 2 sets or Obstacle Course
Friday	Run 15 min (easy pace)	None

WEEK 18	Endurance	Strength
Monday	Team Sport	None
Tuesday	None	Lower Body - (1 set) Power - 1 set
Wednesday	Team Sport	None
Thursday	None	Lower Body - (1 set) Power - 1 set or Obstacle Course
Friday	Team Sport	None

WEEK 19	Endurance	Strength
Monday	Team Sport	None
Tuesday	None	None
Wednesday	Team Sport	None
Thursday	None	None
Friday	Team Sport	Obstacle Course

WEEK 20	Endurance	Strength
Monday	None	Upper/Lower - (2 sets)
Tuesday	Team Sports	None
Wednesday	None	Upper/Lower - (2 sets)
Thursday	Team Sports	Obstacle Course
Friday	None	Upper/Lower - (2 sets)

On week 21, begin the training cycle again, starting at week 1.

The 20-week PT cycle uses a modified form of periodization training. No athlete ca maintain a continuous competitive edge - any attempt to do so will result in an overtrained state where mental and physical abilities will deteriorate.

Interruptions in the training schedule are expected. As a general guideline, for each week during which scheduled PT is not done, the unit should go back 2 weeks from the last week to safely restart the program.

c. 5-Day/6-Week Cardio/Muscular Strength Conditioning Program

MONDAY	TUESDAY	WEDNESDAY	THURSDAY	FRIDAY
WEEK 1				
PFT	Resistance Training (RT)	Aerobic Training (AT) Ability Group Run (ABGR) 3-mi Upper Body Strength Training (UBS)	RT	Hike: 5-mi @ 2hrs/ w/30lbs
WEEK 2 Circuit Training (CT)	RT	ABGR: 4-mi UBS	RT	Unit Run
WEEK 3 ABGR: 4-mi	RT	ABGR: 4-mi	Anaerobic Training (ANT): 6x 100m sprints w/ 45-sec rest	Hike: 11km/6.8-mi <2.75hrs w/40lbs
WEEK 4 RT	ABGR: 5-mi	CT	RT	Unit Run
WEEK 5 ABGR: 5-mi UBS	RT	ABGR: 5-mi	ANT: 8x 100m sprints w/ 45-sec rest	Hike: 13km/8-mi <3.25hrs w/50lbs
WEEK 6 PFT 4-mi	RT	ABGR: 6-mi UBS	RT	Unit Run

d. Conditioning Program "A". Weekly aerobic program to address problems in the PFT and other distance running.

Monday: Distance/Time Run 30-minute run (slower pace than PFT run pace)

Tuesday: Cross training, Conduct cardiovascular training other than running; e.g., circuit training, swimming, or biking

Wednesday: Interval Training

Interval runs on track or known distance course - 4x 400 sprints;
- Increase # of sprints with conditioning
- Increase sprint distance with conditioning

Thursday: Cross Training

Friday: Distance Run - increase duration/speed from Monday's run

e. Conditioning Program "B" (6 months). Program is designed to address PFT/distance running weaknesses, using a long-term (6 months) conditioning period. Run training should be conducted three times per week on non-consecutive days (e.g., Monday/Wednesday/Friday). One of the runs will be an interval run and the remaining two should be distance runs. The following schedule depicts the interval runs for every week for the 6-month period.

INTERVAL RUN TRAINING

WEEK	RUN DIST (YDS:MI)	MAX TIME (MIN:SEC)	WALK DIST: TIME (YDS:SEC)	REPS-RUN/WALK	TOTAL TIME (MIN:SEC)	TOTAL MILES
1,2	220: 1/8	1:00	110:60	11	22:00	2
3,4	330:	1:30	110:60	9	22:30	2.25
5,6	440: 1/4	2:20	110:75	8	26:40	2.5
7,8	660:	4:00	110:90	6	31:00	2.63
9,10	880: 1/2 440: 1/4	5:30 2:10	110:120 110:75	3 3	32:45	2.63
11,12	880: 1/2 440: 1/4	5:20 2:00	110:120 110:75	4 2	35:50	2.81
13,14	880: 1/2 660:	5:20 3:45	110:105 110:75	4 2	38:20	3.06
15,16	880: 1/2 660:	5:00 3:30	110:105 110:75	5 1	36:00	3.19
17,18	880: 1/2	4:45	110:90	6	37:30	3.38
19,20	1 M 4 laps 880: 1/2	9:00 4:30	110:120 110:90	1 4	35:00	3.31
21-24	1 M 4 laps 880: 1/2	8:30 4:15	110:105 110:90	1 4	33:15	3.31

The following chart shows the distance runs (per training session) for every week for the 6-month period. The distance runs should be conducted at 75% of maximum effort.

DISTANCE RUN TABLE

WEEKS	RUN DISTANCE (MILES)
1-4	2.5
5-8	2.75
9-12	3.0
13-16	3.25
17-20	3.5
21-24	4.0

Each PT session should include a substantial warmup, stretching, and cool-down from the Daily 16.

f. Conditioning Program "C" (Single Activity). The "single activity" conditioning schedule provides a 60-minute training session in which the single activity system is used with different events each day. In this program, 5 hours of training are scheduled for the week, and supplemented with integrated unit training each week. A bimonthly conditioning march of 10 miles in 3 hours is suggested to round out this program. Such a program could be used as a transition period (1-4 weeks) between other remedial/conditioning programs.

DAY	TIME	ACTIVITY
Monday	60-min	Running Activity
Tuesday	60-min	Combatives
Wednesday	60-min	Conditioning & Log Drills
Thursday	60-min	Conditioning & Relays
Friday	60-min	Running & Conditioning
Saturday (optional)		Running Activity

g. Conditioning Program "D" (Single Activity). This conditioning program schedule illustrates the use of another "single activity" system for 5.25 hours of training, to include sustaining-type activities for a unit that has passed through both the toughening and slow improvement stages of conditioning. The use of competitive activities is featured to hold interest and provide self-motivation. A bimonthly 10-mile/3-hour march under load will be necessary to round out this program.

DAY	TIME	ACTIVITY
Monday	45-min	Running - Terrain Running
Tuesday	75-min	Team Sports or Ability Group Run/Interval Sets
Wednesday	45-min	Running - Fartlek or Distance Run
Thursday	20-min 40-min	Log Drills Dual Combatives
Friday	90-min	Speed March
Saturday (optional)		Team Sports/Team Contests

h. Conditioning Program "E" (Multi-Activities). This program provides a more comprehensive training regimen of various training periods and activities, and the use of three systems during the training week. In this schedule, 3.5 hours are included for the week.

DAY	TIME	ACTIVITY
Monday	20-min 20-min 20-min	Rifle Drill Combatives Running
Tuesday	30-min	Daily 16 Grass Drills
Wednesday	30-min	Rifle Drill
Thursday	30-min	Circuit Interval Training
Friday	20-min 20-min 20-min	Strength Circuit Conditioning Drills Cross-Country Run
Saturday (optional)	60-min	Warmup Run/Grass Drills

On Monday, a 60-min period is available for PT with three, 20-min segments. Groups will progress through each scheduled activity in turn. On Tuesday, Wednesday, and Thursday, only 30-min periods are available, conducted by the entire unit. As with the Monday schedule, on Friday, the unit should conduct the 20-min segments as subgroups, again progressing through each scheduled activity in turn. Bimonthly conditioning marches will be necessary to augment the program.

APPENDIX D

BODY COMPOSITION PROGRAM (BCP) ASSIGNMENT PROCESS

LETTERHEAD

6100
Office Code
Date

From: Commanding Officer
To: Medical Officer

Subj: BODY COMPOSITION PROGRAM (BCP) EVALUATION ON RANK SNM XXX XX XXX/XXXX

Ref: (a) MCO P6100.12

1. It has been determined that (RANK SNM)____________________________ is not within the Corps' body composition standards.

2. During the semi-annual body composition evaluation conducted on (YYMMDD) ____________________, SNM was found to weigh ________ pounds and has a ________ percent body fat. Per the reference, the Marine's maximum weight is ________ pounds based on a height of ________ inches.

3. SNM received a Physical Fitness Test (PFT) score on (YYMMDD) ________________, which is a ($1^{st}/2^{nd}/3^{rd}$) ________ Class score.

4. SNM was screened and does not meet the (PFT performance/percent body fat) ________________________ criteria and is therefore not eligible for the Physical Performance Evaluation.

5. SNM has been advised that the loss of ______ pounds or _____ percent body fat is required and this reduced weight/body fat percentage must be maintained within the 6-month BCP assignment period.

6. Request a medical evaluation, per the reference, be conducted to determine SNM's medical status for BCP and Remedial Physical Conditioning Program (RPCP) participation.

____
CO's Signature

Copy to:

APPENDIX D

BODY COMPOSITION PROGRAM (BCP) ASSIGNMENT PROCESS
LETTERHEAD

6100
Office Code
Date

FIRST ENDORSEMENT on (identify letter)

From: Medical Officer
To: Commanding Officer

Subj: BODY COMPOSITION PROGRAM (BCP) EVALUATION ON RANK SNM
XXX XX XXXX/XXXX

Ref: (a) MCO P6100.12

1. Returned.

2. Per the reference, it has been determined that (RANK SNM) ____________ is not within body composition standards for the following (initial one of the following):

a. (_____) Due to an underlying cause or associated disease. Attached are the recommended follow-on medical tests/evaluations.

b. (_____) Due to an underlying cause or associated disease. Attached is the prescribed treatment for the condition. The Residential Obesity Program is/is not recommended for this individual.

c. (_____) Not to an underlying cause or associated disease. SNM is fit for participation in a BCP and Remedial Physical Conditioning Program (RPCP). Attached are the prescribed weight and body fat reduction goals and dietary guidelines.

d. (_____) Not to an underlying cause or associated disease. SNM is fit for participation in a BCP and alternative RPCP. Attached are the physical restrictions/limitations imposed, prognosis for return to a full duty status, weight and body fat reduction goals, and dietary guidelines.

____
MO's (ACHCP's) Signature

____
Grade

____
Branch of Service

____
Medical Title

____
Organization

APPENDIX D

BODY COMPOSITION PROGRAM (BCP) ASSIGNMENT PROCESS
LETTERHEAD

6100
Office Code
Date

SECOND ENDORSEMENT on (identify letter)

From: Commanding Officer
To: Individual Marine

Subj: BODY COMPOSITION PROGRAM (BCP) EVALUATION ON RANK SNM
XXX XX XXXX/XXXX

Ref: (a) MCO P6100.12
(b) MCO P1080.40C
(c) MCO P1900.16F

1. Readdressed and forwarded.

2. It has been determined that you are not in adherence with Marine Corps body composition standards and are not eligible for the Physical Performance Evaluation. You are being assigned to a 6-month BCP as a ($1^{st}/2^{nd}$) ________ assignment. You are directed to meet the following weight reduction goal of ______ pounds or a body fat reduction goal of ______ percent per month. Your goal is to lose ______ pounds and ______ percent body fat.

3. Your mandatory participation in the BCP and following Program begins on (YYMMDD) _______________ and ends on (YYMMDD) _______________.

a. (_____) The Unit's Remedial Physical Conditioning Program (RPCP) for 6 months per reference (a).

b. (_____) The alternate RPCP as a result of physical restrictions/ limitations imposed by the Medical Officer per reference (a).

4. You are advised that assistance in meeting and maintaining your weight and body fat reduction goals is available through

____

____

____.

5. Official assignment to the 6-month BCP will be reported on the Unit Diary per reference (b).

6. You are advised that failure to comply with reference (a) may result in administrative separation per paragraph 6206 or 6215 of reference (c) respectively.

____
CO's Signature

____
Unit Diary Number

____
Date

APPENDIX D

BODY COMPOSITION PROGRAM (BCP) ASSIGNMENT PROCESS
LETTERHEAD

6100
Office Code
Date

THIRD ENDORSEMENT on (identify letter)

From: Individual Marine
To: Commanding Officer

Subj: BODY COMPOSITION PROGRAM (BCP) EVALUATION ON RANK SNM
XXX XX XXXX/XXXX

1. Returned with acknowledgments. (SNM must read and initial all paragraphs below.)

2. (_____) I understand I am not in adherence with Marine Corps body composition standards, and it is not due to an underlying cause or associated disease, based on the medical evaluation completed by a Medical Officer (MO).

3. (_____) I understand I am not eligible for the Physical Performance Evaluation and I am being assigned to the BCP, as a (1st/2nd) ________ assignment.

4. (_____) I understand I am being directed to meet the following weight reduction goal of ______ pounds or a body fat reduction goal of ______ percent per month, and my weight goal is ______ pounds and body fat goal is ______ percent.

5. I understand my responsibility to (initial all):

a. (_____) Participate in the Unit's BCP for a total of 6 months.

b. (_____) Participate in the Unit Remedial Physical Conditioning Program (RPCP) or alternate RPCP, as prescribed by the evaluating MO for the duration of my 6-month BCP assignment.

c. (_____) Have my weight and body fat measurements taken and recorded once every 2 weeks by the Command Physical Training Representative.

d. (_____) Follow the dietary regimen prescribed by the evaluating MO.

e. (_____) Complete the MCI Course 3316, "Basic Nutrition." (First assignment to BCP only).

____
Individual Marine's Signature

APPENDIX D

BODY COMPOSITION PROGRAM (BCP) ASSIGNMENT PROCESS
LETTERHEAD

6100
Office Code
Date

FOURTH ENDORSEMENT on (identify letter)

From: Commanding Officer
To: Medical Officer

Subj: BODY COMPOSITION PROGRAM (BCP) EVALUATION ON RANK SNM
XXX XX XXXX/XXXX

1. Readdressed and forwarded for action.

2. SNM has been participating in the BCP and Remedial Physical Conditioning Program for 4 months. A medical reevaluation is requested on SNM, based on progress made towards weight and body fat reduction goals. If weight and body fat reduction goals are not being met or maintained, verify the diagnosis of no underlying cause or associated disease.

3. SNM has lost ______ pound(s) and ______ percent body fat, since being placed to the BCP on (YYMMDD) _______________.

____

CO's Signature

APPENDIX D

BODY COMPOSITION PROGRAM (BCP) ASSIGNMENT PROCESS
LETTERHEAD

6100
Office Code
Date

FIFTH ENDORSEMENT on (identify letter)

From: Medical Officer
To: Commanding Officer

Subj: BODY COMPOSITION PROGRAM (BCP) EVALUATION ON RANK SNM XXX XX XXXX/XXXX

Ref: (a) MCO P6100.12

1. Returned.

2. SNM has been reevaluated, resulting in the following diagnosis/ recommendation: (Initial appropriate paragraph)

a. (_____) SNM's current body composition status is due to an underlying cause or associated disease. Attached are the diagnosis and recommended follow-on actions to be taken per paragraph 3200 of the reference.

b. (_____) SNM's present body composition status is not due to an underlying cause or associated disease. Participation in the BCP and Remedial Physical Conditioning Program should continue until Program assignment expires.

3. Based on the above diagnoses, medical comments and recommendations are attached.

____
MO's (ACHCP's) Signature

____
Grade

____
Branch of Service

____
Medical Title

____
Organization

ENCLOSURE (6)

APPENDIX D

BODY COMPOSITION PROGRAM (BCP) ASSIGNMENT PROCESS
LETTERHEAD

6100
Office Code
Date

SIXTH ENDORSEMENT on (identify letter)

From: Commanding Officer
To: Individual Marine

Subj: BODY COMPOSITION PROGRAM (BCP) EVALUATION ON RANK SNM
XXX XX XXXX/XXXX

Ref: (a) MCO P1080.40C
(b) MCO P1900.16F

1. Readdressed and forwarded.

2. Your (1st/2nd) ________ 6-month assignment to the BCP expired on (YYMMDD) ________________.

3. Based on your progress while assigned to the BCP, as documented in your BCP file, it has been determined that: (Initial one)

a. (_____) You have attained and maintained the Marine Corps' body composition standards, and will be officially removed from the BCP (and Remedial Physical Conditioning Program (RPCP)) on the Unit Diary per reference (a).

(1) (_____) FIRST ASSIGNMENT PARTICIPANTS. If at any time during your Marine Corps career you fail to meet the body composition standards a second time, you may be granted a second BCP (and RPCP) assignment of 6 months with no extension.

(2) (_____) SECOND ASSIGNMENT PARTICIPANTS. If at any time during your Marine Corps career you fail to meet the body composition standards a third time, you will automatically be processed for administrative separation per reference (b).

b. (_____) FIRST ASSIGNMENT PARTICIPANTS ONLY. You have not met the Marine Corps' body composition standards, but have made satisfactory progress during your first 6-month BCP assignment. You have been granted an extension for ______ additional months, and will be required to return to body composition standards upon Program extension expiration on (YYMMDD) ________________. This extension has been entered on the Unit Diary per reference (a).

(1) You will continue your participation in the BCP and RPCP throughout the extension period.

(2) Thirty-days prior to Program expiration, you will be medically reevaluated.

(3) Upon Program expiration, if body composition standards are met, you will be officially removed from the BCP (and RPCP). If body composition standards are not met, you will be processed for administrative separation per paragraph 6215 in reference (b).

SIXTH ENDORSEMENT on (identify letter)

Subj: BODY COMPOSITION PROGRAM (BCP) EVALUATION ON RANK SNM
XXX XX XXXX/XXXX

c. (_____) You have failed to meet the Marine Corps' body composition standards and will receive a 6105 counseling entry on your SRB/OQR's Page 11, and be processed for administrative separation per paragraph 6206 or 6215 in reference (b) respectively.

____
CO's Signature

____
Individual Marine's Signature

____
Unit Diary Number

____
Date

LETTERHEAD

6100
Office Code
Date

SEVENTH ENDORSEMENT on (identify letter)

From: Commanding Officer
To: Individual Marine

Subj: BODY COMPOSITION PROGRAM (BCP) EVALUATION ON RANK SNM
XXX XX XXXX/XXXX

Ref: (a) MCO P1080.40C
(b) MCO P1900.16F

1. Readdressed and forwarded.

2. Your extension to your first 6-month BCP assignment expired on (YYMMDD) ________________.

3. Based on your progress while assigned to the BCP, as documented in your BCP file, it has been determined that: (Initial one)

a. (_____) You have attained and maintained the Marine Corps' body composition standards, and will be officially removed from the BCP (and Remedial Physical Conditioning Program (RPCP)) on the Unit Diary per reference (a). If at any time during your Marine Corps career you fail to meet the body composition standards a second time, you may be granted a second BCP (and RPCP) assignment of 6-months with no extension.

b. (_____) You have failed to meet the Marine Corps' body composition standards and will receive a 6105 counseling entry on your SRB/OQR's Page 11, and be processed for administrative separation per paragraph 6206 or 6215 in reference (b) respectively.

____
CO's Signature

____
Individual Marine's Signature

____
Unit Diary Number

____
Date

APPENDIX D

PHYSICAL PERFORMANCE EVALUATION FLOWCHART

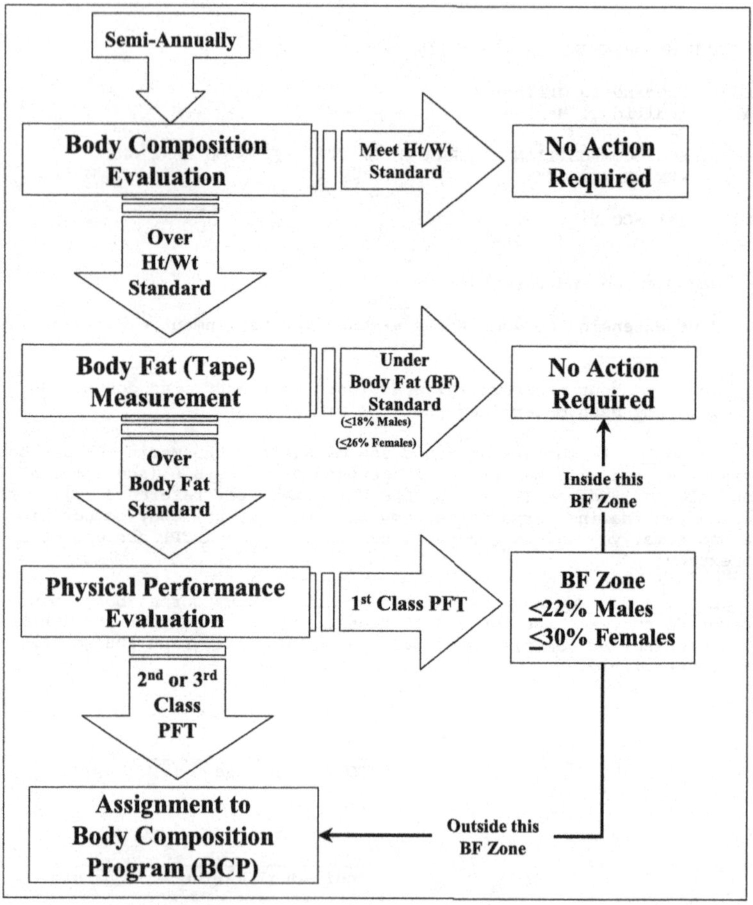

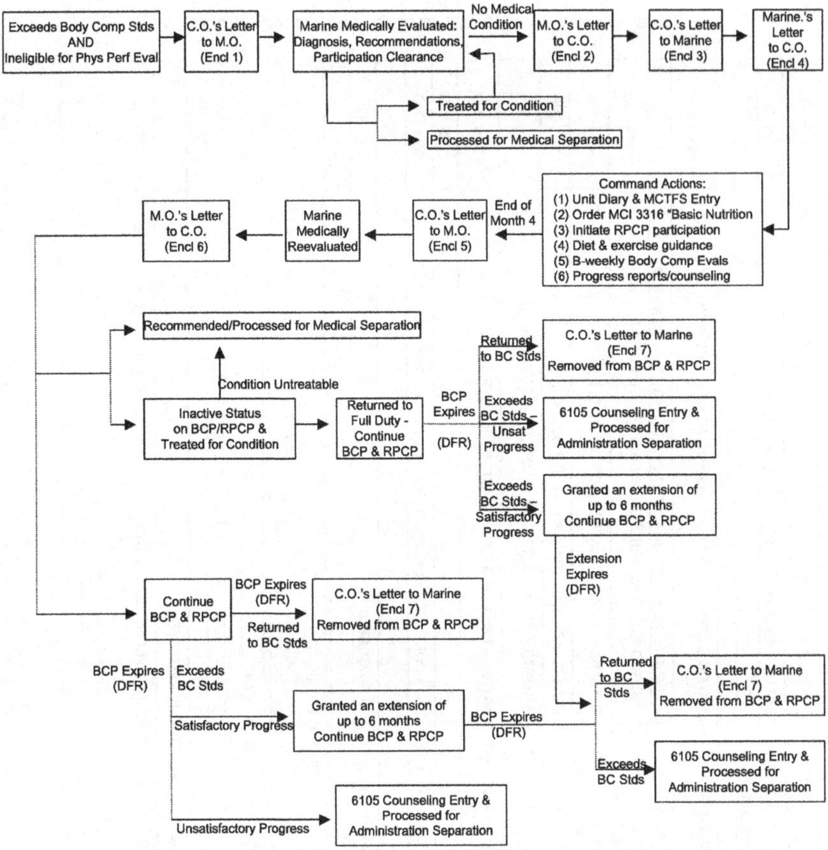

Appendix D: Body Composition Program Assignment Process

NOTES:

(1) Marines assigned to a Body Composition Program (BCP) will also be assigned to the Remedial Physical Conditioning Program (RPCP) for the duration of the 6-month BCP Assignment/Extension.
(2) A Marine on light/limited duty will participate in an alternative RPCP.
(3) 2nd-Assignment Marines, who successfully completed their 1st Assignment and fail to meet BC standards later in their career, will be given 6 months (with no estension) for a 2nd Assignment.
(4) No Third Chances: Marines failing to meet BC standards a third time in their career will be immediately processed for administrative separation, and will informally participate in the BCP & RPCP until discharged. No 3rd Assignment will be given under any circumstances.
(5) Marines will receive a 6105 counseling entry on their SRB/OQR's Page 11 per paragraph 6206 or 6215 of MCO P1900.16F respectively.

APPENDIX E

AGE 46 AND OVER RISK FACTOR SCREENING CHECKLIST

AGE 46 AND OVER RISK FACTOR SCREENING

NAME (LAST, FIRST, M.I.):		SSN:
COMMAND:	SECTION:	PHONE:

SECTION A: (COMPLETED BY THE MARINE)	Circle
(1) ARE YOU NOT IN ACCORDANCE WITH MEDICAL PHYSICAL EXAMINATION REQUIREMENTS FOR YOUR AGE?	YES NO
(2) DATE OF LAST PERIODIC PHYSICAL EXAMINATION:	YES NO
SECTION B: (COMPLETED BY THE MARINE)	
(1) DO YOU HAVE A HISTORY OF HEART DISEASE OR HIGH BLOOD PRESSURE WHICH REQUIRES YOU TO RESTRICT PHYSICAL ACTIVITY OR SEEK MEDICAL TREATMENT?	YES NO
(2) SINCE YOUR LAST PERIODIC PHYSICAL EXAMINATION, SEMI-ANNUAL RISK FACTOR SCREENING, OR PFT, HAS THERE BEEN:	
A. A CHANGE IN YOUR HISTORY OF HEART DISEASE OR HIGH BLOOD PRESSURE WHICH REQUIRED YOU TO RESTRICT PHYSICAL ACTIVITY OR SEEK MEDICAL TREATMENT?	YES NO
B. A CHANGE IN YOUR FAMILY HISTORY SUCH AS IMMEDIATE FAMILY MEMBERS EXPERIENCING A HEART ATTACK OR DIED OF HEART DISEASE BEFORE THEY WERE 45 YEARS OLD?	YES NO
C. A CHANGE IN YOUR INCIDENCE OF DISCOMFORT TO YOUR CHEST, ARMS, OR NECK WHILE EXERCISING?	YES NO
D. AN INCIDENT WHERE YOU WERE PRONE TO HEAT EXHAUSTION/HEAT STROKE, FEELING FAINT, OR FEELING THAT YOU WERE ABOUT TO LOSE CONSCIOUSNESS?	YES NO
E. A CHANGE IN ANY MEDICAL CONDITION (E.G., DIABETES, ASTHMA, BONE OR JOINT DISEASE) WHICH YOU THINK MIGHT LIMIT YOUR PARTICIPATION IN THE PFT?	YES NO
F. A SIGNIFICANT WEIGHT CHANGE?	YES NO
G. A CHANGE IN YOUR SMOKING HABITS SUCH THAT YOU NOW SMOKE ONE OR MORE PACKS OF CIGARETTES DAILY?	YES NO
MARINE'S SIGNATURE : DATE:	

SECTION C: (COMPLETED BY THE MEDICAL OFFICER (M.O.) IF REQUIRED)

DATE OF MEDICAL EVALUATION: ______________________

(1) THE MARINE'S PERIODIC PHYSICAL EXAMINATION IS CURRENT (Y/N)? ________
REMARKS:______________________

(2) THE MARINE ANSWERED 'YES' TO QUESTIONS IN SECTION (B)(Y/N)? ______
REMARKS:______________________

SECTION D: (COMPLETED BY THE MEDICAL OFFICER (M.O.) IF REQUIRED)

(1) MEDICAL RECOMMENDATION: REFERRED FOR FURTHER MEDICAL CONSULTATION (Y/N)? ____ (E.G., PSYCHOLOGY, NUTRITION, INTERNAL MEDICINE, PHYSICAL THERAPY)
REMARKS:______________________

(2) CLEARED FOR PARTICIPATION IN PFT (Y/N)? ____
REMARKS:______________________

SIGNATURE OF M.O. DATE:

PRINTED NAME/RANK/COMMAND/UNIT/PHONE:

APPENDIX F

PHYSICAL FITNESS TEST AND BODY COMPOSITION WORKSHEET

UNIT:

DATE:

MONITOR:

Remarks

* %BF required ONLY if Marine exceeds Ht/Wt standards.

** Semiannual Risk Factor Screening required ONLY if Marine is age 46 and over.

NAME/SSN	RANK	AGE	GENDER M/F	HT (in.)	WT (lbs.)	% BF*	Risk Factor Screening**	Pullups Flx Arm Hang	SCORE	Crunches	SCORE	3-Mile Run	SCORE	PFT TOTAL	Partial PFT TOTAL	CLASS	Phys Perf Eval (Y/N)

APPENDIX G

PHYSICAL FITNESS TEST SCORING TABLE

Sec I: FEMALES

Points	Flexed-Arm Hang	Crunches	3-Mile Run
100	70 sec	100	21:00
99		99	21:10
98	69 sec	98	21:20
97		97	21:30
96	68 sec	96	21:40
95		95	21:50
94	67 sec	94	22:00
93		93	22:10
92	66 sec	92	22:20
91		91	22:30
90	65 sec	90	22:40
89		89	22:50
88	64 sec	88	23:00
87		87	23:10
86	63 sec	86	23:20
85		85	23:30
84	62 sec	84	23:40
83		83	23:50
82	61 sec	82	24:00
81		81	24:10
80	60 sec	80	24:20
79		79	24:30
78	59 sec	78	24:40
77		77	24:50
76	58 sec	76	25:00
75		75	25:10
74	57 sec	74	25:20
73		73	25:30
72	56 sec	72	25:40
71		71	25:50
70	55 sec	70	26:00
69		69	26:10
68	54 sec	68	26:20
67		67	26:30
66	53 sec	66	26:40
65		65	26:50
64	52 sec	64	27:00
63		63	27:10
62	51 sec	62	27:20
61		61	27:30
60	50 sec	60	27:40
59		59	27:50
58	49 sec	58	28:00
57		57	28:10
56	48 sec	56	28:20
55		55	28:30
54	47 sec	54	28:40
53		53	28:50
52	46 sec	52	29:00
51		51	29:10

Points	Flexed-Arm Hang	Crunches	3-Mile Run
50	45 sec	50	29:20
49		49	29:30
48	44 sec	48	29:40
47		47	29:50
46	43 sec	46	30:00
45		45	30:10
44	42 sec	44	30:20
43		43	30:30
42	41 sec	42	30:40
41		41	30:50
40	40 sec	40	31:00
39	39 sec	x	31:10
38	38 sec	x	31:20
37	37 sec	x	31:30
36	36 sec	x	31:40
35	35 sec	x	31:50
34	34 sec	x	32:00
33	33 sec	x	32:10
32	32 sec	x	32:20
31	31 sec	x	32:30
30	30 sec	x	32:40
29	29 sec	x	32:50
28	28 sec	x	33:00
27	27 sec	x	33:10
26	26 sec	x	33:20
25	25 sec	x	33:30
24	24 sec	x	33:40
23	23 sec	x	33:50
22	22 sec	x	34:00
21	21 sec	x	34:10
20	20 sec	x	34:20
19	19 sec	x	34:30
18	18 sec	x	34:40
17	17 sec	x	34:50
16	16 sec	x	35:00
15	15 sec	x	35:10
14	x	x	35:20
13	x	x	35:30
12	x	x	35:40
11	x	x	35:50
10	x	x	36:00
9	x	x	x
8	x	x	x
7	x	x	x
6	x	x	x
5	x	x	x
4	x	x	x
3	x	x	x
2	x	x	x
1	x	x	x

***Round up all values (e.g., 21:01 to 21:09 equals 99 points)**

APPENDIX G

PHYSICAL FITNESS TEST SCORING TABLE

Sec II:	MALES						
Points	Pull-ups	Crunches	3-Mile Run	**Points**	Pull-ups	Crunches	3-Mile Run
100	20	100	18:00	**50**	10	50	26:20
99		99	18:10	**49**		49	26:30
98		98	18:20	**48**		48	26:40
97		97	18:30	**47**		47	26:50
96		96	18:40	**46**		46	27:00
95	19	95	18:50	**45**	9	45	27:10
94		94	19:00	**44**		44	27:20
93		93	19:10	**43**		43	27:30
92		92	19:20	**42**		42	27:40
91		91	19:30	**41**		41	27:50
90	18	90	19:40	**40**	8	40	28:00
89		89	19:50	**39**		x	28:10
88		88	20:00	**38**		x	28:20
87		87	20:10	**37**		x	28:30
86		86	20:20	**36**		x	28:40
85	17	85	20:30	**35**	7	x	28:50
84		84	20:40	**34**		x	29:00
83		83	20:50	**33**		x	29:10
82		82	21:00	**32**		x	29:20
81		81	21:10	**31**		x	29:30
80	16	80	21:20	**30**	6	x	29:40
79		79	21:30	**29**		x	29:50
78		78	21:40	**28**		x	30:00
77		77	21:50	**27**		x	30:10
76		76	22:00	**26**		x	30:20
75	15	75	22:10	**25**	5	x	30:30
74		74	22:20	**24**		x	30:40
73		73	22:30	**23**		x	30:50
72		72	22:40	**22**		x	31:00
71		71	22:50	**21**		x	31:10
70	14	70	23:00	**20**	4	x	31:20
69		69	23:10	**19**		x	31:30
68		68	23:20	**18**		x	31:40
67		67	23:30	**17**		x	31:50
66		66	23:40	**16**		x	32:00
65	13	65	23:50	**15**	3	x	32:10
64		64	24:00	**14**	x	x	32:20
63		63	24:10	**13**	x	x	32:30
62		62	24:20	**12**	x	x	32:40
61		61	24:30	**11**	x	x	32:50
60	12	60	24:40	**10**	x	x	33:00
59		59	24:50	**9**	x	x	x
58		58	25:00	**8**	x	x	x
57		57	25:10	**7**	x	x	x
56		56	25:20	**6**	x	x	x
55	11	55	25:30	**5**	x	x	x
54		54	25:40	**4**	x	x	x
53		53	25:50	**3**	x	x	x
52		52	26:00	**2**	x	X	x
51		51	26:10	**1**	x	x	x

*** Round up all values (e.g., 18:01 to 18:09 equals 99 points)**

APPENDIX H

ALTITUDE COMPENSATION FOR 3.0 MILE PHYSICAL FITNESS RUN CHART

Points	3.0-Mile (Male)	3.0-Mile (Female)	Points	3.0-Mile (Male)	3.0-Mile (Female)	Points	3.0-Mile (Male)	3.0-Mile (Female)
100	19:30	22:30	**65**	25:20	:20	**30**	31:10	34:10
99	199:40	22:40	**64**	25:30	28:30	**29**	31:20	34:20
98	19:50	22:50	**63**	25:40	28:40	**28**	31:30	34:30
97	20:00	23:00	**62**	25:50	28:50	**27**	31:40	34:40
96	20:10	23:10	**61**	26:00	29:00	**26**	31:50	34:50
95	20:20	23:20	**60**	26:10	29:10	**25**	32:00	35:00
94	20:30	23:30	**59**	26:20	29:20	**24**	32:10	35:10
93	20:40	23:40	**58**	26:30	29:30	**23**	32:20	35:20
92	20:50	23:50	**57**	26:40	29:40	**22**	32:30	35:30
91	21:00	24:00	**56**	26:50	29:50	**21**	32:40	35:40
90	21:10	24:10	**55**	27:00	30:00	**20**	32:50	35:50
89	21:20	24:20	**54**	27:10	30:10	**19**	33:00	36:00
88	21:30	24:30	**53**	27:20	30:20	**18**	33:10	36:10
87	21:40	24:40	**52**	27:30	30:30	**17**	33:20	36:20
86	21:50	24:50	**51**	27:40	30:40	**16**	33:30	36:30
85	22:00	25:00	**50**	27:50	30:50	**15**	33:40	36:40
84	22:10	25:10	**49**	28:00	31:00	**14**	33:50	36:50
83	22:20	25:20	**48**	28:10	31:10	**13**	34:00	37:00
82	22:30	25:30	**47**	28:20	31:20	**12**	34:10	37:10
81	22:40	25:40	**46**	28:30	31:30	**11**	34:20	37:20
80	22:50	25:50	**45**	28:40	31:40	**10**	34:30	37:30
79	23:00	26:00	**44**	28:50	31:50	**9**	34:40	37:40
78	23:10	26:10	**43**	29:00	32:00	**8**	34:50	37:50
77	23:20	26:20	**42**	29:10	32:10	**7**	35:00	38:00
76	23:30	26:30	**41**	29:20	32:20	**6**	35:10	38:10
75	23:40	26:40	**40**	29:30	32:30	**5**	35:20	38:20
74	23:50	26:50	**39**	29:40	32:40	**4**	35:30	38:30
73	24:00	27:00	**38**	29:50	32:50	**3**	35:40	38:40
72	24:10	27:10	**37**	30:00	33:00	**2**	35:50	38:50
71	24:20	27:20	**36**	30:10	33:10	**1**	36:00	39:00
70	24:30	27:30	**35**	30:20	33:20			
69	24:40	27:40	**34**	30:30	33:30			
68	24:50	27:50	**33**	30:40	33:40			
67	25:00	28:00	**32**	30:50	33:50			

APPENDIX I

SECTION 1: HEIGHT/WEIGHT STANDARDS

MALES

HEIGHT (Inches)	Maximum Standard (Pounds)	Minimum Standard (Pounds)
58"	132	91
59"	136	94
60"	141	97
61"	146	100
62"	150	104
63"	155	107
64"	160	110
65"	165	114
66"	170	117
67"	176	121
68"	181	125
69"	186	128
70"	192	132
71"	197	136
72"	203	140
73"	208	144
74"	214	148
75"	220	152
76"	226	156
77"	232	160
78"	238	164
79"	244	168
80"	250	173

Max BF%:	18%

FEMALES

HEIGHT (Inches)	Maximum Standard (Pounds)	Minimum Standard (Pounds)
58"	120	91
59"	124	94
60"	128	97
61"	132	100
62"	137	104
63"	141	107
64"	146	110
65"	150	114
66"	155	117
67"	160	121
68"	164	125
69"	169	128
70"	174	132
71"	179	136
72"	184	140
73"	189	144
74"	195	148
75"	200	152
76"	205	156
77"	211	160
78"	216	164
79"	222	168
80"	228	173

Max BF%:	26%

SECTION 2: PERCENT BODY FAT ESTIMATION

PERCENT BODY FAT ESTIMATION FOR MEN

Circumference Value*	Height (in)									
	60.0	60.5	61.0	61.5	62.0	62.5	63.0	63.5	64.0	64.5
13.5	9	9								
14.0	11	11	10	10	10	10	9	9		
14.5	12	12	12	11	11	11	11	10	10	10
15.0	13	13	13	13	12	12	12	12	11	11
15.5	15	14	14	14	14	13	13	13	13	12
16.0	16	16	15	15	15	15	14	14	14	14
16.5	17	17	16	16	16	16	15	15	15	15
17.0	18	18	18	17	17	17	17	16	16	16
17.5	19	19	19	18	18	18	18	17	17	17
18.0	20	20	20	19	19	19	19	18	18	18
18.5	21	21	21	20	20	20	20	19	19	19
19.0	22	22	22	21	21	21	21	20	20	20
19.5	23	23	23	22	22	22	22	21	21	21
20.0	24	24	24	23	23	23	23	22	22	22
20.5	25	25	25	24	24	24	24	23	23	23
21.0	26	26	25	25	25	25	24	24	24	24
21.5	27	27	26	26	26	26	25	25	25	25
22.0	28	27	27	27	27	26	26	26	26	25
22.5	29	28	28	28	28	27	27	27	27	26
23.0	29	29	29	29	28	28	28	28	27	27
23.5	30	30	30	29	29	29	29	28	28	28
24.0	31	31	30	30	30	30	29	29	29	29
24.5	32	31	31	31	31	30	30	30	30	29
25.0	32	32	32	32	31	31	31	31	30	30
25.5	33	33	33	32	32	32	32	31	31	31
26.0	34	34	33	33	33	33	32	32	32	32
26.5	35	34	34	34	34	33	33	33	33	32
27.0	35	35	35	35	34	34	34	34	33	33
27.5	36	36	36	35	35	35	35	34	34	34
28.0	37	36	36	36	36	35	35	35	35	34
28.5			37	37	36	36	36	36	35	35
29.0					37	37	37	36	36	36
29.5								37	37	36
30.0										
30.5										
31.0										
31.5										
32.0										
32.5										
33.0										
33.5										
34.0										
34.5										
35.0										

* Circumference Value = abdomen circumference - neck circumference (in inches)

APPENDIX I

SECTION 2: PERCENT BODY FAT ESTIMATION

PERCENT BODY FAT ESTIMATION FOR MEN

Circumference Value*	Height (in)									
	65.0	65.5	66.0	66.5	67.0	67.5	68.0	68.5	69.0	69.5
13.5										
14.0										
14.5	10	9	9							
15.0	11	11	10	10	10	10	10	9	9	
15.5	12	12	12	11	11	11	11	11	10	10
16.0	13	13	13	13	12	12	12	12	12	11
16.5	14	14	14	14	14	13	13	13	13	12
17.0	16	15	15	15	15	14	14	14	14	14
17.5	17	16	16	16	16	16	15	15	15	15
18.0	18	18	17	17	17	17	16	16	16	16
18.5	19	19	18	18	18	18	17	17	17	17
19.0	20	20	19	19	19	19	18	18	18	18
19.5	21	21	20	20	20	20	19	19	19	19
20.0	22	21	21	21	21	21	20	20	20	20
20.5	23	22	22	22	22	21	21	21	21	21
21.0	24	23	23	23	23	22	22	22	22	21
21.5	24	24	24	24	23	23	23	23	23	22
22.0	25	25	25	25	24	24	24	24	23	23
22.5	26	26	26	25	25	25	25	24	24	24
23.0	27	27	26	26	26	26	26	25	25	25
23.5	28	27	27	27	27	27	26	26	26	26
24.0	28	28	28	28	28	27	27	27	27	26
24.5	29	29	29	29	28	28	28	28	27	27
25.0	30	30	30	29	29	29	29	28	28	28
25.5	31	31	30	30	30	30	29	29	29	29
26.0	31	31	31	31	31	30	30	30	30	29
26.5	32	32	32	32	31	31	31	31	30	30
27.0	33	33	32	32	32	32	32	31	31	31
27.5	34	33	33	33	33	32	32	32	32	32
28.0	34	34	34	34	33	33	33	33	32	32
28.5	35	35	34	34	34	34	34	33	33	33
29.0	36	35	35	35	35	34	34	34	34	34
29.5	36	36	36	36	35	35	35	35	34	34
30.0	37	37	36	36	36	36	35	35	35	35
30.5			37	37	37	36	36	36	36	35
31.0						37	37	36	36	36
31.5									37	37
32.0										
32.5										
33.0										
33.5										
34.0										
34.5										
35.0										

* Circumference Value = abdomen circumference - neck circumference (in inches)

PERCENT BODY FAT ESTIMATION FOR MEN

Circumference Value*	Height (in)									
	70.0	70.5	71.0	71.5	72.0	72.5	73.0	73.5	74.0	74.5
13.5										
14.0										
14.5										
15.0										
15.5	10	10	9	9	9					
16.0	11	11	11	10	10	10	10	10	9	9
16.5	12	12	12	12	11	11	11	11	11	10
17.0	13	13	13	13	13	12	12	12	12	11
17.5	14	14	14	14	14	13	13	13	13	13
18.0	15	15	15	15	15	14	14	14	14	14
18.5	17	16	16	16	16	15	15	15	15	15
19.0	18	17	17	17	17	16	16	16	16	16
19.5	18	18	18	18	18	17	17	17	17	17
20.0	19	19	19	19	19	18	18	18	18	18
20.5	20	20	20	20	19	19	19	19	19	18
21.0	21	21	21	21	20	20	20	20	20	19
21.5	22	22	22	21	21	21	21	21	20	20
22.0	23	23	23	22	22	22	22	22	21	21
22.5	24	24	23	23	23	23	23	22	22	22
23.0	25	24	24	24	24	24	23	23	23	23
23.5	25	25	25	25	25	24	24	24	24	24
24.0	26	26	26	26	25	25	25	25	25	24
24.5	27	27	27	26	26	26	26	26	25	25
25.0	28	28	27	27	27	27	26	26	26	26
25.5	29	28	28	28	28	27	27	27	27	27
26.0	29	29	29	29	28	28	28	28	28	27
26.5	30	30	30	29	29	29	29	28	28	28
27.0	31	30	30	30	30	30	29	29	29	29
27.5	31	31	31	31	30	30	30	30	30	29
28.0	32	32	32	31	31	31	31	31	30	30
28.5	33	32	32	32	32	32	31	31	31	31
29.0	33	33	33	33	32	32	32	32	32	31
29.5	34	34	34	33	33	33	33	32	32	32
30.0	35	34	34	34	34	34	33	33	33	33
30.5	35	35	35	35	34	34	34	34	34	33
31.0	36	36	35	35	35	35	35	34	34	34
31.5	36	36	36	36	36	35	35	35	35	35
32.0	37	37	37	36	36	36	36	36	35	35
32.5				37	37	36	36	36	36	36
33.0							37	37	36	36
33.5										37
34.0										
34.5										
35.0										

* Circumference Value = abdomen circumference - neck circumference (in inches)

SECTION 2: PERCENT BODY FAT ESTIMATION

PERCENT BODY FAT ESTIMATION FOR MEN

Circumference Value*	Height (in)									
	75.0	75.5	76.0	76.5	77.0	77.5	78.0	78.5	79.0	79.5
13.5										
14.0										
14.5										
15.0										
15.5										
16.0										
16.5	10	10	10	10	9	9				
17.0	11	11	11	11	10	10	10	10	10	9
17.5	12	12	12	12	12	11	11	11	11	11
18.0	13	13	13	13	13	12	12	12	12	12
18.5	14	14	14	14	14	13	13	13	13	13
19.0	15	15	15	15	15	14	14	14	14	14
19.5	16	16	16	16	16	15	15	15	15	15
20.0	17	17	17	17	17	16	16	16	16	16
20.5	18	18	18	18	17	17	17	17	17	16
21.0	19	19	19	19	18	18	18	18	18	17
21.5	20	20	20	19	19	19	19	19	18	18
22.0	21	21	20	20	20	20	20	20	19	19
22.5	22	22	21	21	21	21	21	20	20	20
23.0	23	22	22	22	22	22	21	21	21	21
23.5	23	23	23	23	23	22	22	22	22	22
24.0	24	24	24	24	23	23	23	23	23	22
24.5	25	25	25	24	24	24	24	24	23	23
25.0	26	25	25	25	25	25	24	24	24	24
25.5	26	26	26	26	26	25	25	25	25	25
26.0	27	27	27	27	26	26	26	26	26	25
26.5	28	28	27	27	27	27	27	26	26	26
27.0	29	28	28	28	28	28	27	27	27	27
27.5	29	29	29	29	28	28	28	28	28	27
28.0	30	30	29	29	29	29	29	29	28	28
28.5	31	30	30	30	30	30	29	29	29	29
29.0	31	31	31	31	30	30	30	30	30	29
29.5	32	32	31	31	31	31	31	30	30	30
30.0	32	32	32	32	32	31	31	31	31	31
30.5	33	33	33	32	32	32	32	32	32	31
31.0	34	33	33	33	33	33	33	32	32	32
31.5	34	34	34	34	33	33	33	33	33	33
32.0	35	35	34	34	34	34	34	33	33	33
32.5	35	35	35	35	35	34	34	34	34	34
33.0	36	36	36	35	35	35	35	35	34	34
33.5	37	36	36	36	36	36	35	35	35	35
34.0		37	37	37	36	36	36	36	36	35
34.5					37	37	37	36	36	36
35.0								37	37	36

* Circumference Value = abdomen circumference - neck circumference (in inches)

APPENDIX I

SECTION 2: PERCENT BODY FAT ESTIMATION

PERCENT BODY FAT ESTIMATION FOR WOMEN

Circumference Value*	Height (in)									
	58.0	58.5	59.0	59.5	60.0	60.5	61.0	61.5	62.0	62.5
45.0	19									
45.5	20	20	19							
46.0	21	20	20	20	19					
46.5	21	21	21	20	20	20	19	19		
47.0	22	22	22	21	21	20	20	20	19	19
47.5	23	23	22	22	22	21	21	21	20	20
48.0	24	23	23	23	22	22	22	21	21	21
48.5	24	24	24	23	23	23	22	22	22	21
49.0	25	25	24	24	24	23	23	23	22	22
49.5	26	26	25	25	24	24	24	23	23	23
50.0	27	26	26	26	25	25	24	24	24	23
50.5	27	27	27	26	26	26	25	25	25	24
51.0	28	28	27	27	27	26	26	26	25	25
51.5	29	28	28	28	27	27	27	26	26	26
52.0	29	29	29	28	28	28	27	27	27	26
52.5	30	30	29	29	29	28	28	28	27	27
53.0	31	30	30	30	29	29	29	28	28	28
53.5	31	31	31	30	30	30	29	29	29	28
54.0	32	32	31	31	31	30	30	30	29	29
54.5	33	32	32	32	31	31	31	30	30	30
55.0	33	33	33	32	32	32	31	31	31	30
55.5	34	34	33	33	33	32	32	32	31	31
56.0	35	34	34	34	33	33	33	32	32	31
56.5	35	35	35	34	34	34	33	33	32	32
57.0	36	36	35	35	34	34	34	33	33	33
57.5	37	36	36	35	35	35	34	34	34	33
58.0	37	37	36	36	36	35	35	35	34	34
58.5	38	37	37	37	36	36	36	35	35	35
59.0	38	38	38	37	37	37	36	36	36	35
59.5	39	39	38	38	38	37	37	36	36	36
60.0	40	39	39	38	38	38	37	37	37	36
60.5	40	40	39	39	39	38	38	38	37	37
61.0	41	40	40	40	39	39	39	38	38	38
61.5	41	41	41	40	40	40	39	39	38	38
62.0	42	42	41	41	40	40	40	39	39	39

* Circumference Value = waist circumference + hip circumference - neck circumference (in inches)

APPENDIX I

SECTION 2: PERCENT BODY FAT ESTIMATION

PERCENT BODY FAT ESTIMATION FOR WOMEN

Circumference Value*	Height (in)									
	58.0	58.5	59.0	59.5	60.0	60.5	61.0	61.5	62.0	62.5
62.5	42	42	42	41	41	41	40	40	40	39
63.0	43	43	42	42	42	41	41	41	40	40
63.5	44	43	43	42	42	42	41	41	41	40
64.0	44	44	43	43	43	42	42	42	41	41
64.5	45	44	44	44	43	43	43	42	42	42
65.0	45	45	45	44	44	43	43	43	42	42
65.5	46	45	45	45	44	44	44	43	43	43
66.0	46	46	46	45	45	45	44	44	43	43
66.5	47	46	46	46	45	45	45	44	44	44
67.0			47	46	46	46	45	45	45	44
67.5				47	46	46	46	45	45	45
68.0					47	47	46	46	46	45
68.5							47	46	46	46
69.0								47	47	46
69.5										47
70.0										
70.5										
71.0										
71.5										
72.0										
72.5										
73.0										
73.5										
74.0										
74.5										
75.0										
75.5										
76.0										
76.5										
77.0										
77.5										
78.0										
78.5										
79.0										
79.5										

* Circumference Value = waist circumference + hip circumference - neck circumference (in inches)

PERCENT BODY FAT ESTIMATION FOR WOMEN

Circumference Value*	Height (in)									
	63.0	63.5	64.0	64.5	65.0	65.5	66.0	66.5	67.0	67.5
45.0										
45.5										
46.0										
46.5										
47.0										
47.5	19	19								
48.0	20	20	20	19						
48.5	21	21	20	20	20	19				
49.0	22	21	21	21	20	20	20	19	19	
49.5	22	22	22	21	21	21	20	20	20	19
50.0	23	23	22	22	22	21	21	21	21	20
50.5	24	23	23	23	23	22	22	22	21	21
51.0	25	24	24	24	23	23	23	22	22	22
51.5	25	25	25	24	24	24	23	23	23	22
52.0	26	26	25	25	25	24	24	24	23	23
52.5	27	26	26	26	25	25	25	24	24	24
53.0	27	27	27	26	26	26	25	25	25	24
53.5	28	28	27	27	27	26	26	26	25	25
54.0	29	28	28	28	27	27	27	26	26	26
54.5	29	29	29	28	28	28	27	27	27	26
55.0	30	30	29	29	29	28	28	28	27	27
55.5	31	30	30	30	29	29	29	28	28	28
56.0	31	31	30	30	30	30	29	29	29	28
56.5	32	31	31	31	30	30	30	29	29	29
57.0	32	32	32	31	31	31	30	30	30	29
57.5	33	33	32	32	32	31	31	31	30	30
58.0	34	33	33	33	32	32	32	31	31	31
58.5	34	34	34	33	33	33	32	32	32	31
59.0	35	35	34	34	34	33	33	33	32	32
59.5	35	35	35	34	34	34	33	33	33	33
60.0	36	36	35	35	35	34	34	34	33	33
60.5	37	36	36	36	35	35	35	34	34	34
61.0	37	37	37	36	36	36	35	35	35	34
61.5	38	37	37	37	36	36	36	36	35	35
62.0	38	38	38	37	37	37	36	36	36	35

* Circumference Value = waist circumference + hip circumference - neck circumference (in inches)

SECTION 2: PERCENT BODY FAT ESTIMATION

PERCENT BODY FAT ESTIMATION FOR WOMEN

Circumference Value*	Height (in)									
	63.0	63.5	64.0	64.5	65.0	65.5	66.0	66.5	67.0	67.5
62.5	39	39	38	38	38	37	37	37	36	36
63.0	40	39	39	39	38	38	38	37	37	37
63.5	40	40	39	39	39	38	38	38	37	37
64.0	41	40	40	40	39	39	39	38	38	38
64.5	41	41	41	40	40	40	39	39	39	38
65.0	42	41	41	41	40	40	40	39	39	39
65.5	42	42	42	41	41	41	40	40	40	39
66.0	43	42	42	42	41	41	41	41	40	40
66.5	43	43	43	42	42	42	41	41	41	40
67.0	44	44	43	43	43	42	42	42	41	41
67.5	44	44	44	43	43	43	42	42	42	41
68.0	45	45	44	44	44	43	43	43	42	42
68.5	45	45	45	44	44	44	43	43	43	43
69.0	46	46	45	45	45	44	44	44	43	43
69.5	46	46	46	45	45	45	44	44	44	44
70.0	47	47	46	46	46	45	45	45	44	44
70.5			47	46	46	46	46	45	45	45
71.0				47	47	46	46	46	45	45
71.5						47	47	46	46	46
72.0							47	47	46	46
72.5									47	47
73.0										
73.5										
74.0										
74.5										
75.0										
75.5										
76.0										
76.5										
77.0										
77.5										
78.0										
78.5										
79.0										
79.5										

* Circumference Value = waist circumference + hip circumference - neck circumference (in inches)

PERCENT BODY FAT ESTIMATION FOR WOMEN

Circumference Value*	Height (in)									
	68.0	68.5	69.0	69.5	70.0	70.5	71.0	71.5	72.0	72.5
45.0										
45.5										
46.0										
46.5										
47.0										
47.5										
48.0										
48.5										
49.0										
49.5	19									
50.0	20	20	19							
50.5	21	20	20	20	19	19				
51.0	21	21	21	20	20	20	19	19		
51.5	22	22	21	21	21	20	20	20	20	19
52.0	23	22	22	22	21	21	21	21	20	20
52.5	23	23	23	22	22	22	22	21	21	21
53.0	24	24	23	23	23	22	22	22	22	21
53.5	25	24	24	24	23	23	23	23	22	22
54.0	25	25	25	24	24	24	24	23	23	23
54.5	26	26	25	25	25	24	24	24	24	23
55.0	27	26	26	26	25	25	25	25	24	24
55.5	27	27	27	26	26	26	25	25	25	25
56.0	28	28	27	27	27	26	26	26	25	25
56.5	29	28	28	28	27	27	27	26	26	26
57.0	29	29	29	28	28	28	27	27	27	26
57.5	30	29	29	29	29	28	28	28	27	27
58.0	30	30	30	29	29	29	29	28	28	28
58.5	31	31	30	30	30	29	29	29	29	28
59.0	32	31	31	31	30	30	30	29	29	29
59.5	32	32	32	31	31	31	30	30	30	29
60.0	33	32	32	32	32	31	31	31	30	30
60.5	33	33	33	32	32	32	32	31	31	31
61.0	34	34	33	33	33	32	32	32	32	31
61.5	35	34	34	34	33	33	33	32	32	32
62.0	35	35	35	34	34	34	33	33	33	32

* Circumference Value = waist circumference + hip circumference - neck circumference (in inches)

SECTION 2: PERCENT BODY FAT ESTIMATION

PERCENT BODY FAT ESTIMATION FOR WOMEN

Circumference Value*	Height (in)									
	68.0	68.5	69.0	69.5	70.0	70.5	71.0	71.5	72.0	72.5
62.5	36	35	35	35	34	34	34	34	33	33
63.0	36	36	36	35	35	35	34	34	34	34
63.5	37	37	36	36	36	35	35	35	34	34
64.0	37	37	37	36	36	36	36	35	35	35
64.5	38	38	37	37	37	36	36	36	36	35
65.0	38	38	38	38	37	37	37	36	36	36
65.5	39	39	38	38	38	37	37	37	37	36
66.0	40	39	39	39	38	38	38	37	37	37
66.5	40	40	39	39	39	39	38	38	38	37
67.0	41	40	40	40	39	39	39	39	38	38
67.5	41	41	41	40	40	40	39	39	39	38
68.0	42	41	41	41	40	40	40	40	39	39
68.5	42	42	42	41	41	41	40	40	40	39
69.0	43	42	42	42	41	41	41	41	40	40
69.5	43	43	43	42	42	42	41	41	41	41
70.0	44	43	43	43	43	42	42	42	41	41
70.5	44	44	44	43	43	43	42	42	42	42
71.0	45	44	44	44	44	43	43	43	42	42
71.5	45	45	45	44	44	44	43	43	43	43
72.0	46	45	45	45	45	44	44	44	43	43
72.5	46	46	46	45	45	45	44	44	44	44
73.0	47	46	46	46	45	45	45	45	44	44
73.5		47	47	46	46	46	45	45	45	44
74.0				47	46	46	46	46	45	45
74.5					47	47	46	46	46	45
75.0							47	46	46	46
75.5								47	47	46
76.0										47
76.5										
77.0										
77.5										
78.0										
78.5										
79.0										
79.5										

* Circumference Value = waist circumference + hip circumference - neck circumference (in inches)

PERCENT BODY FAT ESTIMATION FOR WOMEN

Circumference Value*	Height (in)									
	73.0	73.5	74.0	74.5	75.0	75.5	76.0	76.5	77.0	77.5
45.0										
45.5										
46.0										
46.5										
47.0										
47.5										
48.0										
48.5										
49.0										
49.5										
50.0										
50.5										
51.0										
51.5										
52.0	20	19	19							
52.5	20	20	20	19	19					
53.0	21	21	20	20	20	20	19	19		
53.5	22	21	21	21	21	20	20	20	19	19
54.0	22	22	22	21	21	21	21	20	20	20
54.5	23	23	22	22	22	22	21	21	21	20
55.0	24	23	23	23	22	22	22	22	21	21
55.5	24	24	24	23	23	23	23	22	22	22
56.0	25	25	24	24	24	23	23	23	23	22
56.5	26	25	25	25	24	24	24	24	23	23
57.0	26	26	26	25	25	25	24	24	24	24
57.5	27	26	26	26	26	25	25	25	25	24
58.0	27	27	27	27	26	26	26	25	25	25
58.5	28	28	27	27	27	27	26	26	26	25
59.0	29	28	28	28	27	27	27	27	26	26
59.5	29	29	29	28	28	28	27	27	27	27
60.0	30	30	29	29	29	28	28	28	28	27
60.5	30	30	30	30	29	29	29	28	28	28
61.0	31	31	30	30	30	30	29	29	29	28
61.5	32	31	31	31	30	30	30	30	29	29
62.0	32	32	32	31	31	31	30	30	30	30

* Circumference Value = waist circumference + hip circumference - neck circumference (in inches)

PERCENT BODY FAT ESTIMATION FOR WOMEN

Circumference Value*	Height (in)									
	73.0	73.5	74.0	74.5	75.0	75.5	76.0	76.5	77.0	77.5
62.5	33	32	32	32	32	31	31	31	30	30
63.0	33	33	33	32	32	32	32	31	31	31
63.5	34	34	33	33	33	32	32	32	32	31
64.0	34	34	34	34	33	33	33	32	32	32
64.5	35	35	34	34	34	33	33	33	33	32
65.0	35	35	35	35	34	34	34	33	33	33
65.5	36	36	35	35	35	35	34	34	34	33
66.0	37	36	36	36	35	35	35	35	34	34
66.5	37	37	37	36	36	36	35	35	35	35
67.0	38	37	37	37	36	36	36	36	35	35
67.5	38	38	38	37	37	37	36	36	36	36
68.0	39	38	38	38	38	37	37	37	36	36
68.5	39	39	39	38	38	38	37	37	37	37
69.0	40	39	39	39	39	38	38	38	37	37
69.5	40	40	40	39	39	39	39	38	38	38
70.0	41	40	40	40	40	39	39	39	38	38
70.5	41	41	41	40	40	40	40	39	39	39
71.0	42	41	41	41	41	40	40	40	39	39
71.5	42	42	42	41	41	41	41	40	40	40
72.0	43	42	42	42	42	41	41	41	40	40
72.5	43	43	43	42	42	42	42	41	41	41
73.0	44	43	43	43	43	42	42	42	41	41
73.5	44	44	44	43	43	43	42	42	42	42
74.0	45	44	44	44	44	43	43	43	42	42
74.5	45	45	45	44	44	44	43	43	43	43
75.0	46	45	45	45	44	44	44	44	43	43
75.5	46	46	46	45	45	45	44	44	44	44
76.0	47	46	46	46	45	45	45	45	44	44
76.5		47	46	46	46	46	45	45	45	44
77.0			47	47	46	46	46	45	45	45
77.5					47	47	46	46	46	45
78.0						47	47	46	46	46
78.5								47	47	46
79.0										47
79.5										

* Circumference Value = waist circumference + hip circumference - neck circumference (in inches)

APPENDIX J

PHYSICAL FITNESS TEST (PFT) ENHANCEMENT TRAINING

1. Background. The PFT is an excellent instrument for measuring a Marine's general fitness level, as it relates to upper body strength and cardio-vascular endurance. Although the PFT is not specifically designed to evaluate a Marine's level of combat fitness, it does provide a baseline for potential success in meeting the physical demands of combat. This Appendix provides an abridged guide for PFT training that both reviews and applies the general principles of strength and conditioning for each of the PFT events.

2. General Strength and Conditioning Principles

a. Specificity. The main benefit of specificity is specific training develops neuromuscular efficiency. Movement that is more efficient means more repetitions or faster times. This principle states that to become good at a particular movement, you must practice that particular movement. For example, a pistol shooter does not become good at rapid fire by shooting continual precision strings of fire. Specificity states the shooter would be better served by specific training because specific training yields specific gains.

(1) Pullup/Flexed Arm Hang. There are many exercises and machines that can enhance upper body strength, but there is no better exercise for improving pullup or the flexed-arm hang performance than the act of doing or practicing them.

(2) Crunch. The various abdominal strengthening devices (machines) and routines that focus on the mid- to lower abdominal muscles will improve crunch performance in most cases. However, doing or practicing crunches is the best method for improving crunch performance.

(3) 3.0 Mile Run. Exercise machines such as stair climbers, stationary bicycles, elliptical cross trainers, and rowing machines can enhance a Marine's cardiovascular endurance. They cannot, however replace running as the activity that will best improve running performance.

b. Overload. The acronym "SAID" stands for "Specific Adaptations to Imposed Demands," and describes what happens when the muscular and cardiovascular systems are overloaded or "pushed." When a Marine does an activity to a point of fatigue, muscle systems that have been exerted will become stronger and better able to handle the increased exercise stimulus. Proper rest and nutrition are important cofactors in ensuring the benefits of overloading are realized. It is important for the Marine and the commander to understand that improper overloading could have a detrimental effect on the desired result.

(1) Pullup/Flexed-Arm Hang. Routines that require completing set(s) of pullups to the point of exhaustion are worthwhile. Form and slowed execution of the pullup should be maintained in order to facilitate momentary muscular fatigue. Executing pullups under weight (e.g., with a pack or flak jacket on) and varying the width of the grip (e.g., narrow to wide grip) are excellent training methods. Training for pullups in this manner will help ensure that a Marine will be able to do proper pullups in an all-out effort on test day.

(2) Crunch. Applying the principle of overload in preparing for the crunch requires doing two sets of crunches until momentary muscular fatigue. Tracking the amount of "complete" crunches a Marine can do until the point of fatigue, in a certain time period, will provide excellent feedback on the effectiveness of the overload principle. A second way to train for the crunch event would be for a Marine to do as many crunches as he can, without stopping, until complete muscle failure. Focusing on using the abdominal muscles (keeping them tight throughout the exercise), and executing them in slow motion is an excellent way to intensify the training. Track the number of proper repetitions executed to determine effectiveness.

(3) 3.0 Mile Run. Interval training done once or twice a week can enhance a Marine's running performance. Interval training sessions take less time, but require a higher level of intensity to run the shorter distances. The 400-meter and 800-meter courses are the best distances for increasing performance for the 3.0 mile run. If the distance interval training were longer than 800 meters, the Marine would not be able to run at the targeted speed. If the distance is shorter than 400 meters, the Marine cannot reach the targeted level of intensity needed for endurance training.

(a) To calculate the time an individual Marine should run the 400/800-meter distances during an interval training session is as follows:

1 Determine the most recent 3.0 mile run time;

2 Divide that time by 3. This gives the mile-split time for the run;

3 Divide the mile-split time by 2. Subtract that time by 10-20 seconds. This gives the time the 800-meter intervals should be run.

4 To determine the 400-meter interval time, divide the mile-split time by 4. Then subtract 5-10 seconds. This is the time the 400-meter intervals should be run.

(b) Calculating rest in between each running exercise is very important. Complete recovery in between the intervals lessens the benefit of overload, but not enough rest is also detrimental. Incomplete recovery in between intervals should be the goal. The work/rest ratio for 800-meter intervals should be 1:1. In other words, active rest in the form of light jogging or walking (not stopping) should be equal to the time that was run. For example if a 4:00 minute interval time was run, then rest should be 4:00 minutes. The work/rest ratio for 400-meter intervals should be 1:2. Using the same type of rest as in 800-meter interval session, a 1 min 40 second 400-meter interval would be followed by a 3 min 20 second rest period.

1 Example: A Marine wants to do interval sessions to improve the 22:00 he ran in the last PFT.

a Divide 22:00 by 3 (7:20);

b Divide 7:20 by 2 to determine the pace for an 800-meter split time for a 7:20 mile pace. (3:40);

c Subtract 10-20 seconds to determine the pace for 800-meter intervals. (3:20-3:40);

d The rest between these 800 repeats should be 3:20-3:40.

e Determine the pace for a 400-meter interval by dividing 7:20 by 4 to determine the pace for a 400-meter split time for 7:20 mile pace (1:50). Subtract 5-10 seconds to determine pace for 400-meter intervals. (1:40-1:45);

f The rest between these 400 repeats should be 3:20-3:30.

2 A sample (beginner's) interval workout:

- Daily 16 Warmup
- 800 meter jog
- 800-rest-800-rest-400-rest-400
- 800 meter jog
- Daily 16 Cool-Down

(c) The total mileage for this workout is 2.5 miles with 1.5 miles being speed-work. Many Marines and recreational runners do intervals that are much too fast for their goal time. Keeping the predetermined goal pace should be the goal for this workout. A Marine should not feel totally exhausted following this training session. As a Marine gets more comfortable with this training, more intervals can be added.

(d) Interval training should be combined with other types of running, during the weekly training schedule, for the training to be effective. Longer distance runs along with tempo running (faster running but not at race pace), combined with interval running will improve most Marines' 3.0 mile run times.

3. Summary

a. Specificity and overload are two of the most important strength and conditioning principles. Other principles include:

(1) Regularity - Strength and cardiovascular training should be done 3-4 times a week.

(2) Variety - Varying training regimes periodically prevents boredom.

(3) Recovery - Rest is physiologically necessary for growth and repair.

(4) Balance - A good training plan should exercise all major muscle groups, and include warmup and cool-down phases.

(5) Progression - Ensures that systems are continually challenged.

b. There are many ways in which Marines can improve their performance on the PFT. By applying the strength and conditioning principles listed above, training can be enhanced to improve not only PFT performance but increase a Marines' physical readiness for combat.

APPENDIX K

LIGHT/LIMITED DUTY ASSIGNMENT PROCESS

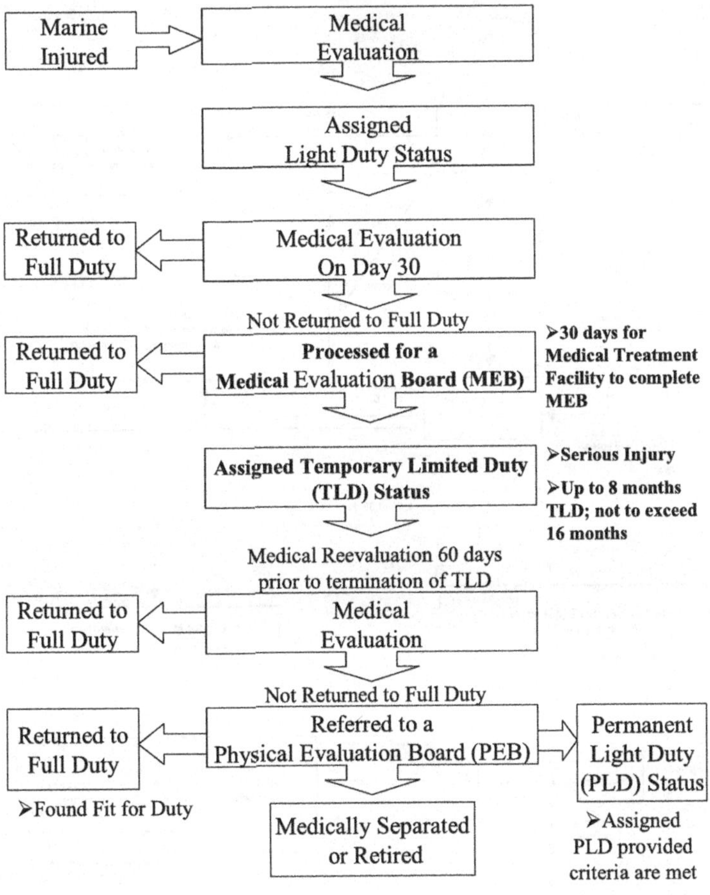

NOTE:

- Marines on light or limited duty (TLD & PLD) are required to maintain Height/Weight or Body Fat Standards regardless of duty status. Marines not maintaining standards can be assigned to the Body Composition Program.
- Marines assigned to light or limited duty (TLD & PLD) who are medically restricted from completing all three events of a PFT will be screened for participation in a Partial PFT – 2 of the 3 events.

APPENDIX L

PREGNANCY PFT AND BODY COMPOSITION PROGRAM FLOWCHART

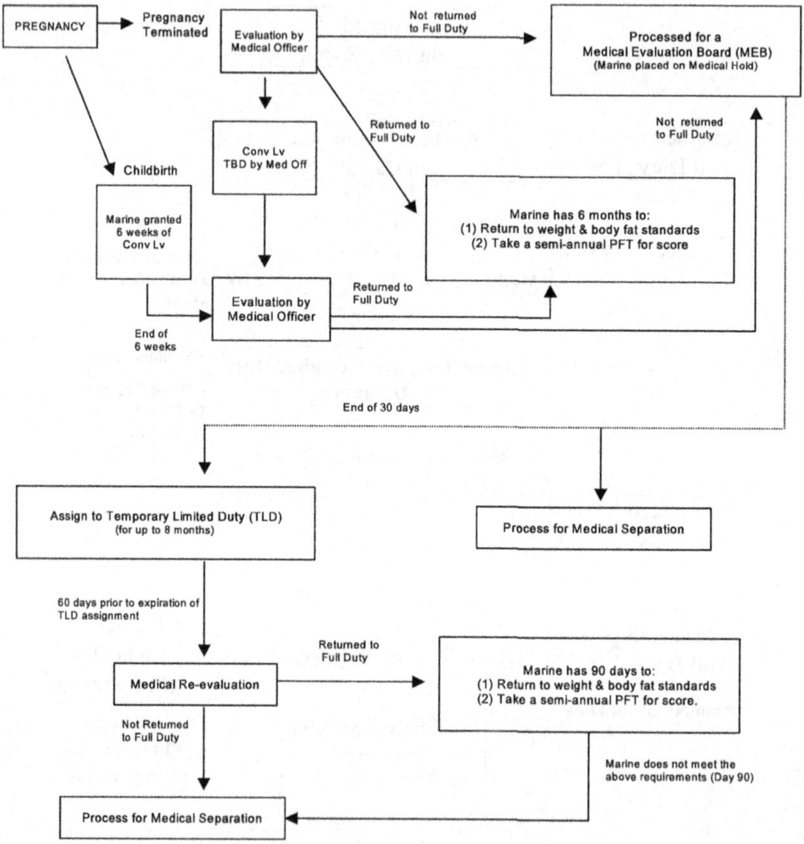

NOTE: Marines assigned to a Body Composition Program (BCP) PRIOR to pregnancy will go into an "inactive status". Once returned to full duty, Marine will return to an active status and will have 6 months from that day to return to weight & body fat standards, regardless on time spent on Program previously.

Made in United States
North Haven, CT
04 November 2026

21774368R00147